THE THEORY OF SPINORS

THE THEORY OF SPINORS

ÉLIE CARTAN

Foreword by
Raymond Streater

Dover Publications, Inc.
New York

This Dover edition, first published in 1981, is an unabridged republication of the complete English translation first published by Hermann of Paris in 1966. The work first appeared in French in 1937 as *Leçons sur la théorie des spineurs* (2 volumes), which was printed from Élie Cartan's lectures, gathered and arranged by André Mercier.

International Standard Book Number: 0-486-64070-1
Library of Congress Catalog Card Number: 80-69677

Manufactured in the United States of America
Dover Publications, Inc., 31 East 2nd Street, Mineola, N.Y. 11501

FOREWORD

Elie Cartan was one of the founders of the modern theory of Lie groups, a subject of central importance in mathematics, and also one with many applications to physics.

In these notes, Cartan describes the representations orthogonal groups, either with real or complex parameters, including reflections, and also the related groups with indefinite metric. The treatment emphasizes the geometric point of view, and would be regarded as elementary by post-war mathematicians. The result is a detailed, explicit treatise that can be understood by the reader (whether he is a trained mathematician or not).

To keep the subject elementary, the author has stated without proof the general theorems of Weyl, and Peter and Weyl, on complete reducibility and existence of representations of a general class of groups. These results are explicitly demonstrated for the orthogonal groups.

Concerning the applications to physics, the rotation and Lorentz groups are naturally the most important. In fact, Cartan shows how to derive the "Dirac" equation for any group, and extends the equation to general relativity. He does not, however, show the relation of the equation to the corresponding inhomogeneous groups; this was discovered only later. The lectures touch on the relation of Clifford algebras to the orthogonal groups, which has become important in recent work on the theory of seniority in atomic spectra, and also contains enough material on the group O(3, 3) to enable one to start on the study of the $\tilde{U}(4)$ theory of Salam, Delboirgo and Strathdee, and other recent theories of strong interactions.

INTRODUCTION

Spinors were first used under that name, by physicists, in the field of Quantum Mechanics. In their most general mathematical form, spinors were discovered in 1913 by the author of this work, in his investigations on the linear representations of simple groups*; they provide a linear representation of the group of rotations in a space with any number n of dimensions, each spinor having 2^ν components where $n = 2\nu + 1$ or 2ν. Spinors in four-dimensional space occur in Dirac's famous equations for the electron, the four wave functions being nothing other than the components of a spinor. Numerous papers have been published on spinors in general; Hermann Weyl and Richard Brauer have recently published an excellent paper† which may be considered as fundamental, although several of the results obtained are very briefly indicated in the paper referred to above. In an unpublished course given at Princeton University, O. Veblen has made a very interesting study of spinors from another point of view. But in almost all these works, spinors are introduced in a purely formal manner, without any intuitive geometrical significance; and it is this absence of geometrical meaning which has made the attempts to extend Dirac's equations to general relativity so complicated.

One of the principal aims of this work is to develop the theory of spinors systematically by giving a purely geometrical definition of these mathematical entities: because of this geometrical origin, the matrices used by physicists in Quantum Mechanics appear of their own accord, and we can grasp the profound origin of the property, possessed by Clifford algebras, of representing rotations in space having any number of dimensions. Finally this geometrical origin makes it very easy to introduce spinors into Riemannian

* E. Cartan, "Les groupes projectifs qui ne laissent invariante aucune multiplicité plane" *Bull. Soc. Math. France*, **41** 1913, 53–96.

† R. Brauer and H. Weyl, "Spinors in *n* dimensions" *Am. J. Math.*, **57**, 1935, 425–449.

geometry, and particularly to apply the idea of parallel transport to these geometrical entities. The difficulties which have been encountered in this respect—difficulties which are *insurmountable* if classical techniques of Riemannian geometry are used—can be explained. These classical techniques are applicable to vectors and to ordinary tensors, which, besides their metric character, possess a purely affine character; but they cannot be applied to spinors which have metric but not affine characteristics.

This course of lectures is divided into two parts. The first is devoted to generalities on the group of rotations in n-dimensional space and on the linear representations of groups, and to the theory of spinors in three-dimensional space, and finally, linear representations of the group of rotations in that space are examined. The importance of these representations in Quantum Mechanics is well known; the infinitesimal element treatment is used for their determination, as this method requires the least preliminary discussion; despite its great interest, the transcendental method of H. Weyl, based on the theory of characters, has been completely omitted.

The second part is devoted to the theory of spinors in spaces of any number of dimensions, and particularly in the space of special relativity; the linear representations of the Lorentz group are referred to, as well as the theory of spinors in Riemannian geometry.

ELIE CARTAN

TABLE OF CONTENTS

PART II

Spinors in Space of $n > 3$ Dimensions
Spinors in Riemannian Geometry

THE
THEORY OF SPINORS

PART I

Spinors in three-dimensional space Linear representations of the group of rotations

CHAPTER I

n-DIMENSIONAL EUCLIDEAN SPACE; ROTATIONS AND REVERSALS

I. EUCLIDEAN SPACE

1. Definition; Vectors

Points in n-dimensional Euclidean space may be defined as sets of n numbers $(x_1, x_2, \ldots, x_n)$, the square of the distance from a point (x) to the origin $(0, 0, \ldots, 0)$ being given by the *fundamental form*

$$\Phi \equiv x_1^2 + x_2^2 + \ldots + x_n^2; \tag{1}$$

this expression also represents the scalar square or the square of the length of the *vector* $\mathbf{x}$ drawn from the origin to the point (x); the n quantities x_i are called the *components* of this vector. The co-ordinates x_i may be arbitrary complex numbers, and it is then said that the Euclidean space is complex; but if they are all real, we have real Euclidean space. In the real domain there also exists pseudo-Euclidean spaces, each corresponding to a real non-positive definite fundamental form

$$\Phi \equiv x_1^2 + x_2^2 + \ldots + x_{n-h}^2 - x_{n-h+1}^2 - \ldots - x_n^2; \tag{2}$$

we shall assume, without any loss of generality, that $n - h \geqslant h$.

In real spaces, we are led to consider vectors whose components are not all real; such vectors are said to be complex.

A vector is said to be *isotropic* if its scalar square is zero, that is to say if its components make the fundamental form equal to zero. A vector in real or complex Euclidean space is said to be a unit vector if its scalar square equals 1. In real pseudo-Euclidean space whose fundamental form is not positive definite, a distinction is made between real *space-like vectors* which have positive fundamental forms, and real *time-like vectors* which have negative fundamental forms; a space-like unit vector has a fundamental form equal to $+1$, for a time-like unit vector it equals -1.

If we consider two vectors $\mathbf{x}$, $\mathbf{y}$, and the vector $\mathbf{x} + \lambda\mathbf{y}$, where λ is a given parameter, that is to say the vector with components $x_i + \lambda y_i$, the scalar square of this vector is

$$\mathbf{x}^2 + \lambda^2\mathbf{y}^2 + 2\lambda\,\mathbf{x}\,.\,\mathbf{y},$$

where $\mathbf{x}\,.\,\mathbf{y}$ is defined as the sum $x_1y_1 + x_2y_2 + \ldots + x_ny_n$. This sum is the *scalar product* of the two vectors; in the case of a pseudo-Euclidean space, the scalar product is

$$x_1y_1 + x_2y_2 + \cdots + x_{n-h}y_{n-h} - x_{n-h+1}y_{n-h+1} - \cdots - x_ny_n.$$

Two vectors are said to be *orthogonal*, or perpendicular to each other, if their scalar product is zero; an isotropic vector is perpendicular to itself. The space of the vectors orthogonal to a given vector is a hyperplane of $n - 1$ dimensions (defined by a linear equation in the co-ordinates).

2. Cartesian frames of reference

The n vectors $\mathbf{e}_1, \mathbf{e}_2, \mathbf{e}_3, \ldots, \mathbf{e}_n$, whose components are all zero except one which is equal to 1, constitute a *basis*, in the sense that every vector $\mathbf{x}$ is a linear combination $x_1\mathbf{e}_1 + x_2\mathbf{e}_2 + \cdots + x_n\mathbf{e}_n$ of these n vectors. These basis vectors are orthogonal in pairs; they constitute what we shall call an orthogonal Cartesian frame of reference.

More generally, let us take n linearly independent vectors, $\boldsymbol{\eta}_1, \boldsymbol{\eta}_2, \ldots, \boldsymbol{\eta}_n$ i.e., such that there is no system of numbers $\lambda_1, \lambda_2, \ldots, \lambda_n$ which are not all zero and which would make the vector $\lambda_1\boldsymbol{\eta}_2 + \lambda_2\boldsymbol{\eta}_2 + \cdots + \lambda_n\boldsymbol{\eta}_n$ identically zero. Every vector $\mathbf{x}$ can then be uniquely expressed in the form $u^1\boldsymbol{\eta}_1 + u^2\boldsymbol{\eta}_2 + \cdots + u^n\boldsymbol{\eta}_n$. The scalar square of this vector is

$$u^iu^j\boldsymbol{\eta}_i\,.\,\boldsymbol{\eta}_j,$$

a formula in which, following Einstein's convention, the summation sign has been suppressed: the index i and the index j take, successively and independently of one another, all the values $1, 2, \ldots, n$. If we write

$$g_{ij} = g_{ji} = \boldsymbol{\eta}_i\,.\,\boldsymbol{\eta}_j, \tag{3}$$

the fundamental form becomes

$$\Phi \equiv g_{ij}u^iu^j. \tag{4}$$

We say further that the set of vectors $\boldsymbol{\eta}_1, \boldsymbol{\eta}_2, \ldots, \boldsymbol{\eta}_n$ forms a basis, or that they constitute a Cartesian frame of reference. We shall consider only basis vectors drawn from the same origin.

Conversely, let us try to determine whether a quadratic expression given *a priori* is capable, in a given Euclidean space, of representing the fundamental form by a suitable choice of frame of reference. We shall naturally assume that the variables and the coefficients are complex if the space is complex, or real if the space is real. We shall apply a classical theorem in the theory of quadratic forms.

3. Reduction of a quadratic form to a sum of squares

THEOREM. *Every quadratic form can be reduced to a sum of squares by a linear transformation of the variables.*

The proof which we shall give applies equally well to the real domain as to the complex domain. Let us first assume that one of the coefficients g_{11}, $g_{22}, g_{33}, \ldots, g_{nn}$ is not zero, g_{11} for example. Consider the form

$$\Phi_1 \equiv \Phi - \frac{1}{g_{11}}(g_{11}u^1 + g_{12}u^2 + \cdots + g_{1n}u^n)^2;$$

this no longer contains the variable u^1; if we put

$$y_1 = g_{11}u^1 + g_{12}u^2 + \cdots + g_{1n}u^n,$$

we then have

$$\Phi \equiv \frac{1}{g_{11}} y_1^2 + \Phi_1,$$

where Φ_1 is a quadratic form in the $n - 1$ variables $u^2, u^3, \ldots, u^n$. If on the contrary all the coefficients $g_{11}, g_{22}, \ldots, g_{nn}$ are zero, one at least of the other coefficients, let us say g_{12}, must be non-zero. In this case, put

$$\Phi_2 \equiv \Phi - \frac{2}{g_{12}}(g_{21}u^1 + g_{23}u^3 + \cdots + g_{2n}u^n)(g_{12}u^2 + g_{13}u^3 + \cdots + g_{1n}u^n);$$

the form Φ_2 no longer contains the variables u^1 and u^2; if we now put

$$y_1 + y_2 = g_{21}u^1 + g_{23}u^3 + \cdots + g_{2n}u^n,$$

$$y_1 - y_2 = g_{12}u^2 + g_{13}u^3 + \cdots + g_{1n}u^n,$$

which define the new variables y_1 and y_2, we have

$$\Phi \equiv \frac{2}{g_{12}}(y_1^2 - y_2^2) + \Phi_2;$$

here Φ_2 is a quadratic form in the $n - 2$ variables $u^3, u^4, \ldots, u^n$.

If we now treat Φ_1 and Φ_2 as we have treated Φ, then Φ will be reduced step by step to a sum of squares of independent linear functions of the original variables, each square being multiplied by a constant factor. In the real domain, the operations will not introduce any imaginary element.

Let

$$\Phi \equiv \alpha_1 y_1^2 + \alpha_2 y_2^2 + \cdots + \alpha_\nu y_\nu^2 \qquad (\nu \leqslant n).$$

If we are in the complex domain, we shall take $y_i\sqrt{\alpha_i}$ as new variables and Φ will then be reduced to a sum of squares. If we are in the real domain, we must distinguish the positive coefficients α_i from the negative coefficients α_i; we can still, by taking $y_i\sqrt{\pm\alpha_i}$ as new variables, reduce Φ to the form

$$\Phi \equiv y_1^2 + y_2^2 + \cdots + y_p^2 - y_{p+1}^2 - \cdots - y_q^2.$$

4. Sylvester's law of inertia

In the real domain, the number of positive squares and the number of negative squares are independent of the method of reduction.

Let us assume that there are two different reductions

$$\Phi \equiv y_1^2 + y_2^2 + \cdots + y_p^2 - z_1^2 - z_2^2 - \cdots - z_q^2,$$

$$\Phi \equiv v_1^2 + v_2^2 + \cdots + v_{p'}^2 - w_1^2 - w_2^2 - \cdots - w_{q'}^2;$$

the linear forms y_i and z_j are all independent, and so are the linear forms v_i and w_j. Let us assume that $p \neq p'$, for example that $p < p'$. We have the identity

$$y_1^2 + y_2^2 + \cdots + y_p^2 + w_1^2 + w_2^2 + \cdots + w_{q'}^2$$

$$\equiv v_1^2 + v_2^2 + \cdots + v_{p'}^2 + z_1^2 + z_2^2 + \cdots + z_q^2.$$

Consider the $p + q'$ linear equations

$$y_1 = 0, y_2 = 0, \ldots, y_p = 0, w_1 = 0, w_2 = 0, \ldots w_{q'} = 0;$$

since $p + q' < p' + q' \leqslant n$, these equations have at least one solution for which the unknowns $u^1, u^2, \ldots, u^n$ are not all zero; for this solution we have also

$$v_1 = 0, v_2 = 0, \ldots v_{p'} = 0, z_1 = 0, z_2 = 0_1 \ldots, z_q = 0;$$

it is thus possible to satisfy $p' + q'$ independent equations

$$v_1 = 0, v_2 = 0, \ldots, v_{p'} = 0, \quad w_1 = 0, w_2 = 0, \ldots, w_{q'} = 0$$

by solving a system with a smaller number $(p + q')$ of equations, which is absurd. Therefore $p = p'$, $q = q'$.

Let us add that if the given form Φ in $u^1, u^2, \ldots, u^n$ *is not degenerate*, i.e., if the n-forms $(\partial\Phi/\partial u^1), (\partial\Phi/\partial u^2), \ldots, (\partial\Phi/\partial u^n)$ are independent, or equivalently, if the *discriminant*

$$g = \begin{vmatrix} g_{11} & g_{12} \cdots g_{1n} \\ \cdot & \cdot \quad \cdots\cdot \\ g_{n1} & g_{n2} \cdots g_{nn} \end{vmatrix} \tag{5}$$

of the form is non-zero, the number, $p + q$, of independent squares obtained in the reduction to a sum of squares equals n. For otherwise the n derivatives $\partial\Phi/\partial u^i$, which are linear combinations of the $p + q < n$ forms $y_1, y_2, \ldots, y_p, z_1, \ldots, z_q$ would not be independent.

5. Complex domain

Having proved these theorems, let us return to our problem. We take as our starting point a *non-degenerate* quadratic form (4). In the complex domain there exist n independent linear forms

$$x_i = a_{ik}u^k \ (i = 1, 2, \ldots, n)$$

such that

$$\Phi \equiv x_1^2 + x_2^2 + \cdots + x_n^2.$$

If then, in complex Euclidean space, we consider the vectors $\boldsymbol{\eta}_k$ with components $(a_{1k}, a_{2k}, \ldots, a_{nk})$ $(k = 1, 2, \ldots, n)$, these n vectors are independent; the vector $u^k\boldsymbol{\eta}_k$ has for its ith component $a_{ik}u^k = x_i$, and its scalar square $x_1^2 + x_2^2 + \cdots + x_n^2$ is equal to the given quadratic form. The original fundamental form of the space may therefore, by a suitable choice of basis vectors, be brought into the quadratic form $g_{ij}u^iu^j$ arbitrarily given.

In the real domain the reasoning is the same but the quadratic form $g_{ij}u^iu^j$ should: (1) not be degenerate; (2) be reducible to a sum of $n - h$ positive squares and h negative squares, where h is a given integer.

6. Contravariant and covariant components

Let us suppose Euclidean space to be referred to any Cartesian reference frame and let

$$\Phi \equiv g_{ij}u^iu^j$$

be its fundamental form. The scalar square of the vector $\mathbf{x} + \lambda\mathbf{y}$ is equal to

$$\Phi(\mathbf{x}) + \lambda^2\Phi(\mathbf{y}) + \lambda\left(y^i\frac{\partial\Phi}{\partial x^i}\right);$$

from this result it follows that *the scalar product of two vectors* $\mathbf{x}$ *and* $\mathbf{y}$ *is*

$$\tfrac{1}{2}y^i\frac{\partial\Phi}{\partial x^i} = g_{ij}y^ix^j;$$

in particular, the geometrical significance of the coefficients is rediscovered if $\mathbf{x}$ and $\mathbf{y}$ are taken to be two basis vectors.

The *covariant components* of a vector $\mathbf{x}$ are the scalar products $\mathbf{x}\cdot\mathbf{e}_1$, $\mathbf{x}\cdot\mathbf{e}_2, \ldots, \mathbf{x}\cdot\mathbf{e}_n$. They are denoted by $x_1, x_2, \ldots, x_n$. We have therefore

$$x_i = \mathbf{x}\cdot\mathbf{e}_i = g_{ik}x^k; \qquad \mathbf{x}\cdot\mathbf{y} = g_{ij}x^iy^j = x^iy_i = x_iy^i. \tag{6}$$

The ordinary components x^i are said to be *contravariant.* By solving (6) we pass from covariant components to contravariant components; this gives

$$x^i = g^{ik}x_k, \tag{7}$$

where g^{ij} is the ratio of the minor of g_{ij} to the discriminant of the fundamental form. In the case of the form (1), we have $x_i = x^i$.

II. ROTATIONS AND REVERSALS*

7. Definition

The members of the set of linear transformations of the co-ordinates which leave the fundamental form invariant are called "rotations" or "reversals" (about the origin). If under an operation of this sort two vectors $\mathbf{x}$ and $\mathbf{y}$ are

* "Rotations et retournements": the terms "proper and improper orthogonal transformations" are often used.

transformed into $\mathbf{x}'$ and $\mathbf{y}'$, then $\mathbf{x} + \lambda\mathbf{y}$ is transformed into $\mathbf{x}' + \lambda\mathbf{y}'$. The magnitudes of all vectors, and the scalar product of any pair of vectors, are unaltered. In a pseudo-Euclidean real space space-like vectors are transformed into space-like vectors, and time-like vectors into time-like vectors.

Every operation of this type transforms all rectangular frames of reference into rectangular frames of reference. Conversely let $(\mathbf{e}_1, \mathbf{e}_2, \ldots, \mathbf{e}_n)$ and $(\boldsymbol{\eta}_1, \boldsymbol{\eta}_2, \ldots, \boldsymbol{\eta}_n)$ be two rectangular reference frames, and let us suppose that $\boldsymbol{\eta}_i = a_i^k \mathbf{e}_k$. Refer the points of space to the first set of basis vectors and let the corresponding co-ordinates be x^i. Then the transformation

$$(x^i)' = a_k^i x^k$$

has the effect of transforming $\mathbf{e}_i$ into the vector with components $(a_i^1, a_i^2, \ldots, a_i^n)$ which is the vector $\boldsymbol{\eta}_i$; this transformation will leave the fundamental form Φ unaltered, since the scalar square $\Phi(\mathbf{x}')$ of the vector $(x^i)'\mathbf{e}_i = a_k^i x^k \mathbf{e}_i = x^k \boldsymbol{\eta}_k$ is equal to $\Phi(\mathbf{x})$.

8. Transformation determinants

We now prove the following theorem.

THEOREM. *The determinant of the transformation which defines a rotation or a reversal equals* $+1$ *or* -1.

It is sufficient to give a proof for the complex domain, because any real rotation in real Euclidean space can be regarded as a special rotation in complex space.

Take a rectangular frame of reference and let the equations

$$x_i' = a_{ik} x_k \quad (i = 1, 2, \ldots, n) \tag{8}$$

define the transformation. The invariance of the fundamental form requires that

$$a_{1i}^2 + a_{2i}^2 + \cdots + a_{ni}^2 = 1 \; (i = 1, 2, \ldots, n)$$

and

$$a_{1i}a_{1j} + a_{2i}a_{2j} + \cdots + a_{ni}a_{nj} = 0 \;\; (i \neq j).$$

If we use these relations in the expression for the product of the determinant of the transformation by its transpose, the resulting determinant has zero elements everywhere except on the leading diagonal where each of the elements equals one.

If we start from any frame of reference, which corresponds to taking co-ordinates u^k given in terms of the x_i by

$$x_i = \alpha_{ik} u^k,$$

and if the transformation of the u^i corresponding to (8) is given by

$$(u^i)' = b_k^i u^k, \tag{9}$$

then

$$\alpha_{ik} b_h^k u^h = a_{ik} \alpha_{kh} u^h \;\; (i, h = 1, 2, \ldots, n)$$

i.e.,

$$\alpha_{ik}b^k_h = a_{ik}\alpha_{kh} = c_{ik} \text{ (say)}$$

Denoting the determinants formed from c_{ij}, b^i_j, $a_{ij}\alpha_{ij}$ by c, b, a and α, then

$$c = \alpha b = a\alpha,$$

thus $b = a$; that is $b = \pm 1$.

The operation given by a linear transformation of determinant $+1$ is called a "rotation", that given by a linear transformation with determinant -1 is called a "reversal".

9. Reflections

Given a hyperplane π containing the origin, then an operation which makes any point x correspond to its mirror-image in the hyperplane π is called a "reflection", i.e., x' lies on the perpendicular from x to the hyperplane produced to a distance equal to the length of the perpendicular. Take any Cartesian reference frame, and let the hyperplane have the equation

$$a_i x^i = 0.$$

Then $\mathbf{x}'$ is defined by the two conditions

(i) The vector $\mathbf{x}' - \mathbf{x}$ is perpendicular to the hyperplane π;

(ii) The point defined by $\frac{1}{2}(\mathbf{x}' + \mathbf{x})$ lies on the hyperplane.

The equation of the hyperplane shows that the vector $\mathbf{x}$ to any point on it is orthogonal to the vector with covariant components a_i, i.e.,

$$(x^i)' - x^i = \lambda a^i \quad \text{or} \quad (x^i)' = x^i + \lambda a^i.$$

Then

$$a_i(2x^i + \lambda a^i) = 0 \quad \text{or} \quad \lambda = -2\frac{a_i x^i}{a_i a^i}.$$

Thus the operation is possible, provided that $a_i a^i \neq 0$, i.e., provided that the direction perpendicular to the hyperplane is not an isotropic direction, and apart from this case

$$(x^i)' = x^i - 2a^i \frac{a_k x^k}{a_k a^k}.$$

It is easy to verify the invariance of magnitude; $x'_i(x^i)' = x_i x^i$.

We shall use the term *reflection* quite generally for the operation defined above; any reflection is associated with a hyperplane, or more simply with a non-isotropic vector $\mathbf{a}$ (which can be taken to be of unit length).

In dealing with real spaces with non-positive definite forms it is important to distinguish between reflections associated with real space-like and real time-like vectors. We shall call these reflections *space reflections* and *time reflections* respectively.

Any reflection is a reversal. To verify this, it is sufficient to take a particular Cartesian reference frame, e.g., if the first basis vector is the vector associated with the reflection, the equation defining the reflection becomes

$$(u^1)' = -u^1, \quad (u^2)' = u^2, \ldots, (u^n)' = u^n,$$

which obviously gives a determinant equal to -1.

10. Factorisation of a rotation into a product of reflections

We shall now prove the following theorem, which holds for both real and complex domains.

Any rotation is the product of an even number $\leqslant n$ of reflections, any reversal is the product of an odd number $\leqslant n$ of reflections.

The fact that an even number of reflections gives a rotation, whilst an odd number gives a reversal, is a consequence of the fact that a reflection is itself a reversal.

The theorem obviously holds for $n = 1$; we shall assume that it is true for spaces of $1, 2, \ldots, n-1$ dimensions, and show that it must also hold in n-dimensional space.

The theorem is easily proved for the special case of rotations and reversals which leave a non-isotropic vector invariant. Take this vector as $\boldsymbol{\eta}_1$ and choose $n-1$ other basic vectors all lying in the hyperplane π orthogonal to $\boldsymbol{\eta}_1$. The fundamental form becomes

$$\Phi = g_{11}(u^1)^2 + g_{ij}u^iu^j = g_{11}(u^1)^2 + \Psi;$$

the quadratic form Ψ is non-degenerate in $n-1$ variables. The rotation (or reversal) considered will leave the hyperplane invariant, i.e., it leaves u^1 unaltered and transforms the variables $u^2, u^3, \ldots, u^n$ amongst themselves in such a way as to leave the form Ψ unchanged; it is thus determined completely by a rotation or a reversal of the Euclidian space of $n-1$ dimensions which forms the hyperplane π. By hypothesis this can be factorised into reflections associated with vectors in π and the number of reflections required is, at the most, $n-1$.

We come now to the general case. Let $\mathbf{a}$ be any non-isotropic vector, and suppose it is transformed into $\mathbf{a}'$ by the operation under consideration. If the vector $\mathbf{a}' - \mathbf{a}$ is non-isotropic, the reflection associated with it obviously transforms $\mathbf{a}$ into $\mathbf{a}'$; the given operation can thus be considered as this reflection followed by another operation which leaves the non-isotropic vector $\mathbf{a}'$ unchanged, i.e., it results from a series of, at the most, n reflections.

The argument breaks down if for all vectors $\mathbf{x}$, the vector $\mathbf{x}' - \mathbf{x}$ is isotropic, where $\mathbf{x}'$ is the transform of $\mathbf{x}$. We now consider this case; the vectors $\mathbf{x}' - \mathbf{x}$ form a subspace in the sense that $(\mathbf{x}' - \mathbf{x}) + \lambda(\mathbf{y}' - \mathbf{y})$ belongs to it if both $\mathbf{x}' - \mathbf{x}$ and $\mathbf{y}' - \mathbf{y}$ do so also. If this subspace has dimension p, we shall refer to it as a "p-plane". Take p vectors $\boldsymbol{\eta}_1, \boldsymbol{\eta}_2, \ldots, \boldsymbol{\eta}_p$ as basis for this p-plane. The linear space orthogonal to these p vectors is defined by p independent linear equations; it has dimension $n-p$ but it also includes each of the (isotropic) vectors $\boldsymbol{\eta}_i$. We can take as basis for this space the vectors $\boldsymbol{\eta}_1, \boldsymbol{\eta}_2, \ldots, \boldsymbol{\eta}_p$ together with $n-2p$ other independent vectors $\boldsymbol{\xi}_1, \boldsymbol{\xi}_2, \ldots \boldsymbol{\xi}_{n-2p}$*.

* We note that the dimension p of any isotropic subspace must satisfy $n - 2p \geqslant 0$ i.e., $p \leqslant n/2$.

Finally take p further vectors $\boldsymbol{\theta}_1, \boldsymbol{\theta}_2, \ldots, \boldsymbol{\theta}_p$, to complete a basis for the whole space.

If $\mathbf{x}$ and $\mathbf{y}$ are any two vectors, then

$$\mathbf{x}' - \mathbf{x} = a^i\boldsymbol{\eta}_i \quad \text{and} \quad \mathbf{y}' - \mathbf{y} = b^k\boldsymbol{\eta}_k.$$

Then, it follows from the equality of $\mathbf{x}' \cdot \mathbf{y}'$ and $\mathbf{x} \cdot \mathbf{y}$ that

$$\mathbf{x} \cdot (b^k\boldsymbol{\eta}_k) = -\mathbf{y} \cdot (a^k\boldsymbol{\eta}_k).$$

Putting $\mathbf{y}$ equal to any of the vectors $\boldsymbol{\eta}_i$ or $\boldsymbol{\xi}_j$ the right-hand side of the equality is zero, i.e., the vector $b^k\boldsymbol{\eta}_k$ is orthogonal to any vector $\mathbf{x}$; this necessitates that $b^k\boldsymbol{\eta}_k$ is zero, i.e., for these special vectors $\mathbf{y}$, $\mathbf{y}' = \mathbf{y}$. The vectors $\boldsymbol{\eta}_i$ and $\boldsymbol{\xi}_j$ are therefore invariant under the transformation we are considering.

An arbitrary vector can be written as $u^i\boldsymbol{\eta}_i + v^j\boldsymbol{\xi}_j + w^k\boldsymbol{\theta}_k$; its fundamental form is

$$\Phi \equiv u^iw^k\boldsymbol{\eta}_i \cdot \boldsymbol{\theta}_k + v^jv^k\boldsymbol{\xi}_j \cdot \boldsymbol{\xi}_k + v^jw^k\boldsymbol{\xi}_j \cdot \boldsymbol{\theta}_k + w^jw^k\boldsymbol{\theta}_j \cdot \boldsymbol{\theta}_k.$$

Since this is non-degenerate, the coefficients of $u^1, u^2, \ldots, u^p$ are independent linear expressions in $w^1, w^2, \ldots, w^p$. We can choose the basis vectors $\boldsymbol{\theta}_k$ so that

$$\boldsymbol{\eta}_i \cdot \boldsymbol{\theta}_j = \begin{cases} 1 & \text{if} \quad i = j, \\ 0 & \text{if} \quad i \neq j. \end{cases}$$

We can then arrange for $\partial\Phi/\partial w^i$, which equals u^i plus a linear expression in the v^j and the w^k, to reduce to u^i. This requires that each vector $\boldsymbol{\xi}_j$, $\boldsymbol{\theta}_k$ be modified by adding a linear combination of the $\boldsymbol{\eta}_i$. Then

$$\Phi \equiv \Sigma u^iw^i + \gamma_{jk}v^jv^k,$$

where the second term is not degenerate. The vectors $\boldsymbol{\xi}_j$ are invariant under the transformation (rotation or reversal) under consideration; thus if $n > 2p$, we have the case of invariance of a non-isotropic vector for which the theorem has been proved. There remains the case $n = 2p$. Then

$$\boldsymbol{\eta}_i' = \boldsymbol{\eta}_i, \ \boldsymbol{\theta}_i' = \boldsymbol{\theta}_i + \beta_{ik}\boldsymbol{\eta}_k;$$

Using the equalities

$$\boldsymbol{\eta}_i \cdot \boldsymbol{\theta}_i = 1, \ \boldsymbol{\eta}_i \cdot \boldsymbol{\theta}_j = 0 \quad (i \neq j)$$

in the equation expressing the invariance of the products $\boldsymbol{\theta}_i \cdot \boldsymbol{\theta}_j$, it follows that $\beta_{ij} + \beta_{ji} = 0$.

We can simplify the preceding formulae for the transform of an arbitrary linear combination $w^k\boldsymbol{\theta}_k$ of the vectors $\boldsymbol{\theta}_k$: it is

$$w^k\boldsymbol{\theta}_k' = w^k\boldsymbol{\theta}_k + w^k\beta_{kh}\boldsymbol{\eta}_h.$$

Now the sum $w^k\beta_{kh}\boldsymbol{\eta}_h$ will be invariant under any change in the basis $\boldsymbol{\eta}_i$ accompanied by the correlated change in the basis $\boldsymbol{\theta}_k$; under such a change of basis the $\boldsymbol{\eta}_i$ and the w_i transform in the same way (because both Σu^iw^i — the fundamental form — and $\Sigma u^i\boldsymbol{\eta}_i$ are invariant). But it is always possible, by the same transformation of both sets of variables, to reduce an alternating bilinear form to the canonical form, which in the present case is

$$(w^1\eta_2 - w^2\eta_1) + (w^3\eta_4 - w^4\eta_3) + \cdots + (w^{2q-1}\eta_{2q} - w^{2q}\eta_{2q-1})$$

Thus by an appropriate change of basis the transformation equations become

$$\theta_1' = \theta_1 + \eta_2, \theta_2' = \theta_2 - \eta_1, \ldots, \theta_{2q-1}' = \theta_{2q-1} - \eta_{2q}, \theta_{2q}' = \theta_{2q} + \eta_{2q-1},$$

where we must have $p = 2q$ otherwise the vector $\theta_p + \eta_p$ would be non-isotropic and invariant. The space splits into sets of four-dimensional subspaces spanned by $(\eta_1, \eta_2, \theta_1, \theta_2)$, $(\eta_3, \eta_4, \theta_3, \eta_4)$ etc., each of which has a non-degenerate fundamental form which is invariant under the transformation under consideration. It will be sufficient to show that the transformation in each of the four-dimensional subspaces can be expressed by at the most four reflections. Consider the first one. The fundamental form is

$$u^1w^1 + u^2w^2$$

and the rotation (or reversal) is described by

$$(u^1)' = u^1 + w^2, \quad (u^2)' = u^2 - w^1, \quad (w^1)' = w^1, \quad (w^2)' = w^2.$$

The determinant is $+1$, i.e., the transformation is a rotation. It can be shown without difficulty that it results from four successive reflections associated with the non-isotropic vectors

$$\eta_2 + \theta_2, \quad \alpha\eta_2 + \theta_2, \quad \eta_1 + \alpha\eta_2 + \left(1 - \frac{1}{\alpha}\right)\theta_2, \quad \eta_1 + \eta_2 + \left(1 - \frac{1}{\alpha}\right)\theta_2,$$

where α is a number not equal to zero or 1. Thus the theorem has been proved.

In the case of real Euclidean spaces, the above proof involves only real vectors, the only provision being that α is taken as being real.

11. Continuity of the group of rotations

We shall show that in complex space and in real space with positive definite fundamental form the group of rotations is continuous*. This means that any rotation can be connected to the identity transformation by a continuous series of rotations. It is only necessary to prove the result for a rotation arising from two reflections. Suppose two reflections are defined by unit vectors **a** and **b**; provided that $n \geqslant 3$ there is at least one unit vector **c** orthogonal to both **a** and **b**; consider the continuous series of rotations resulting from pairs of successive reflections associated with the unit vectors

$$\mathbf{a}' = \mathbf{a}\cos t + \mathbf{c}\sin t, \qquad \mathbf{b}' = \mathbf{b}\cos t + \mathbf{c}\sin t,$$

where the real parameter t varies from 0 to $\pi/2$; for $t = 0$ this reduces to the given rotation; for $t = \pi/2$ the rotation is the product of the same reflection (that associated with **c**) taken twice in succession, i.e., the identity rotation.

* By saying that rotations form a group we assert the following two properties of rotations: (i) The product of any two rotations is a rotation. (ii) To each rotation there corresponds an inverse rotation.

THEOREM. *In complex Euclidean space, and in real Euclidean space with positive definite fundamental form, the set of rotations (real in the latter case) forms a continuous group.*

12. Proper rotations and improper rotations

We shall show that in a real pseudo-Euclidean space (with non-positive definite fundamental form), the group of rotations is not continuous, but consists of two disjoint sets, which we call the set of *proper rotations* and the set of *improper rotations.*

LEMMA 1. *Two space-like vectors can always be connected by a continuous series of space-like unit vectors.*

Take as fundamental form

$$F \equiv x_1^2 + x_2^2 + \cdots + x_{n-h}^2 - x_{n-h+1}^2 - \cdots - x_n^2.$$

A real space-like vector $\mathbf{x}$ is defined by n real numbers, of which the first $n - h$ can be considered as the components of a vector $\mathbf{u}$ in the real Euclidian space E_{n-h}, and the last h as the components of a vector $\mathbf{v}$ in the real Euclidian space E_h. If $\mathbf{x}$ is a unit vector, then $\mathbf{u}^2 - \mathbf{v}^2 = 1$; we can write

$$\mathbf{u} = \mathbf{a} \cosh \alpha, \qquad \mathbf{v} = \mathbf{b} \sinh \alpha$$

where α is a real number and $\mathbf{a}$ and $\mathbf{b}$ are two unit vectors in E_{n-h} and E_h. Another space-like unit vector $\mathbf{x}'$ can be defined in a similar manner by a real number α' and two real unit vectors $\mathbf{a}'$ and $\mathbf{b}'$. We can pass in a continuous manner from $\mathbf{x}$ to $\mathbf{x}'$:

1. Keep $\mathbf{a}$ and $\mathbf{b}$ fixed and let α vary continuously to α';

2. Keep $\mathbf{b}$ fixed, then $\mathbf{a}$ can be connected to $\mathbf{a}'$ by a continuous sequence of real unit vectors in E_{n-h}.

3. Keep $\mathbf{a}'$ fixed, then $\mathbf{b}$ can be connected to $\mathbf{b}'$ by a continuous sequence of real unit vectors in E_h.

The same Lemma obviously applies to two time-like unit vectors.

LEMMA 2. *The rotation resulting from two space-like (or two time-like) vectors can be connected to the identity rotation by a continuous series of rotations.*

Suppose the rotation results from the reflections associated with two space-like unit vectors $\mathbf{u}$ and $\mathbf{v}$, and let $\mathbf{w}_t$ be a continuous sequence of space-like unit vectors connecting $\mathbf{u}$ to $\mathbf{v}$; the rotations resulting from the reflection associated with $\mathbf{w}_t$ and $\mathbf{v}$ form a continuous series joining the given rotations to the identity rotation.

LEMMA 3. *For any rotation (or reversal) the Jacobian of $x'_1, x'_2, \ldots, x'_{n-h}$ with respect to $x_1, x_2, \ldots, x_{n-h}$ is non-zero.*

Suppose the result is not true. Then there will be a set of values $x_1 = a_1$, $x_2 = a_2, \ldots, x_{n-h} = a_{n-h}$ not all zero, which when substituted in the expressions for $x'_1, x'_2, \ldots, x'_{n-h}$, as linear functions of $x_1, x_2, \ldots, x_n$, makes the total contributions from the terms in these first $n - h$ components zero. The transform of $(a_1, a_2, \ldots, a_{n-h}, 0, \ldots, 0)$, which is a space-like vector, has its first $n - h$ components $x'_1, x'_2, \ldots, x'_{n-h}$ all zero, i.e., it is a time-like vector, which is absurd.

We now come to the proof of the theorem. From lemma 3 it follows that any pair of rotations which can be connected by a continuous series of rotations give the same sign to the Jacobian Δ of $x'_1, x'_2, \ldots x'_{n-h}$ with respect to $x_1, x_2, \ldots, x_{n-h}$, since in passing from one rotation to the other, Δ varies continuously but is never zero. By lemma 2, a rotation consisting of an even number of space-like reflections and an even number of time-like reflections gives the same sign to Δ as does the identity rotation, i.e., plus. A rotation resulting from one space-like reflection and one time-like reflection can by lemma 1 be continuously connected by a sequence of rotations to the rotation resulting from the reflections associated with the vectors $\mathbf{e}_1$ and $\mathbf{e}_n$; thus it makes Δ the same sign as the latter rotation, i.e., minus. A rotation derived from an odd number of space-like reflections and an odd number of time-like reflections can be similarly reduced to this last case by a continuous sequence of rotations. We thus have the following theorem:

THEOREM. *In a real pseudo-Euclidean space the group of rotations consists of two disjoint sets; the first consists of the group of proper rotations, which result from an even number of space-like reflections and an even number of time-like reflections; the second consists of the set of improper rotations, which result from an odd number of space-like reflections and an odd number of time-like reflections; this second set is not a group.*

Proper and improper rotations can be recognised by the sign of the Jacobian of $x'_1, x'_2, \ldots, x'_{n-h}$ with respect to $x_1, x_2, \ldots, x_{n-h}$, or also by the Jacobian of $x'_{n-h+1}, \ldots, x'_n$ with respect to $x_{n-h+1}, \ldots, x_n$.

We consider below *proper reversals*; these result from an odd number of space-like reflections and an even number of time-like reflections. They are characterised by the property that the Jacobian of $x'_{n-h+1}, \ldots, x'_n$ with respect to $x_{n-h+1}, \ldots, x_n$ is positive (*invariance of the direction of time*).

13. Case of spaces with $h = 1$

In these spaces all proper rotations and proper reversals are products of space-like reflections. Thus suppose that $\mathbf{x}$ is a time-like unit vector and $\mathbf{x}'$ is its transform; then taking $\mathbf{x}$ as the basis vector $\mathbf{e}_n$, since the rotation or reversal is proper, the coefficient of x_n in the expression for x'_n is positive, and since $\mathbf{x}'$ is a time-like unit vector, the component x'_n must be greater than one; the scalar product $\mathbf{x} \cdot \mathbf{x}'$ is thus less than -1. The scalar square of the vector $\mathbf{x}' - \mathbf{x}$ is $-2\mathbf{x} \cdot \mathbf{x}' - 2 > 0$; it is thus a space-like vector. Therefore one can go from $\mathbf{x}$ to $\mathbf{x}'$ by a space reflection and to obtain the rotation (or reversal) under consideration, it is only necessary to operate in the real Euclidean space of positive definite form which is orthogonal to $\mathbf{x}'$; this requires at most $n - 1$ space reflections.

THEOREM. *In a space whose fundamental form can be reduced to the sum of $n - 1$ positive squares and one negative square, all proper rotations and all proper reversals can be obtained from at most n space reflections.*

III. MULTIVECTORS

14. Volume of the hyper-parallelepiped constructed on n vectors

Given n vectors $\mathbf{x}, \mathbf{y}, \mathbf{z}, \ldots, \mathbf{t}$ in a definite order, all belonging to a Euclidean space E_n which is referred to any Cartesian reference frame, consider the determinant formed from the contravariant components of these vectors

$$\Delta = \begin{vmatrix} x^1 & x^2 \ldots x^n \\ y^1 & y^2 \ldots y^n \\ \ldots & \ldots \\ t^1 & t^2 \ldots t^n \end{vmatrix};$$

since a rotation can be expressed by a linear substitution of determinant 1, and a reversal by a linear substitution of determinant -1, it follows that the value of Δ is unaltered, except perhaps for its sign, when all the vectors undergo the same rotation or reversal. A similar result holds for the determinant Δ' formed from the covariant components of the vectors; the value of Δ' is the product of Δ by the determinant of the transformation from contravariant to covariant components, that is, by g.

By forming the product

$$\Delta\Delta' = \begin{vmatrix} x^1 & x^2 \ldots x^n \\ y^1 & y^2 \ldots y^n \\ \ldots & \ldots \\ t^1 & t^2 \ldots t^n \end{vmatrix} \begin{vmatrix} x_1 & x_2 \ldots x_n \\ y_1 & y_2 \ldots y_n \\ \ldots & \ldots \\ t_1 & t_2 \ldots t_n \end{vmatrix},$$

it is seen that

$$\Delta\Delta' = \begin{vmatrix} \mathbf{x}^2 & \mathbf{x}\cdot\mathbf{y} \ldots \mathbf{x}\cdot\mathbf{t} \\ \mathbf{y}\cdot\mathbf{x} & \mathbf{y}^2 \ldots \mathbf{y}\cdot\mathbf{t} \\ \ldots & \ldots \\ \mathbf{t}\cdot\mathbf{x} & \mathbf{t}\cdot\mathbf{y} \qquad \mathbf{t}^2 \end{vmatrix};$$

in this latter determinant the given vectors occur only as scalar products. For $n = 3$ this determinant equals the square of the volume of the parallelepiped formed from the three vectors. It is natural to apply the name hyper-volume to the square root of the determinant for any value of n. Thus we have

$$V^2 = \Delta\Delta' = g\Delta^2,$$

i.e.,

$$V = \sqrt{g}\Delta = \frac{1}{\sqrt{g}}\Delta'.$$

In a real pseudo-Euclidean space it is more convenient to write

$$V = \sqrt{|g|}\Delta = \frac{1}{\sqrt{|g|}}\Delta'.$$

The sign of Δ gives the orientation of the n-ad formed from the n vectors; this orientation is the same as that of the reference frame if Δ is positive.

15. Multivectors

Consider now a system of p vectors $\mathbf{x}, \mathbf{y}, \ldots, \mathbf{z}$ taken in a definite order. We say that they form a p-vector, and it is convenient to say that two p-vectors are equal if both sets of p-vectors span the same linear manifold (of dimension p) and if the volumes of the parallelepipeds, constructed in this manifold on each of the p-vectors, is the same, and they have the same orientation. We shall show that a p-vector is completely defined by the determinants of order p which can be formed from the array of the contravariant (or covariant) components of the p vectors. The set of ${}^nC_p$ determinants will be said to be the *components* of the p-vector.

Let $\mathbf{t}$ be a variable vector; for this vector to belong to the same linear manifold as the given vectors it is necessary and sufficient that all determinants of order $p + 1$ constructed from the array

$$\begin{pmatrix} x^1 & x^2 \ldots x^n \\ y^1 & y^2 \ldots y^n \\ \cdots & \cdots \\ z^1 & z^2 \ldots z^n \\ t^1 & t^2 \ldots t^n \end{pmatrix}$$

be zero. Denote the determinant formed from the first p rows and the columns $i_1, i_2, \ldots, i_p$ by $P^{i_1 i_2 \ldots i_p}$; then the equations

$$t^{i_1}P^{i_2 i_3 \ldots i_{p+1}} - t^{i_2}P^{i_1 i_3 \ldots i_{p+1}} + \cdots (-1)^p t^{i_{p+1}}P^{i_1 i_2 \ldots i_p} = 0$$

hold.

Thus if two p-vectors with components P and Q span the same linear manifold it is necessary and sufficient that their corresponding components be proportional.

Assume this condition is satisfied and $Q^{i_1 i_2 \ldots i_p} = P^{i_1 i_2 \ldots i_p}$; the algebraic ratio of the two p-vectors, a quantity which is unchanged by projection, is equal to the ratio of any pair of their components; these components represent the volumes of their projections on the various co-ordinate manifolds.

Finally we note that the square of the magnitude of a p-vector (V^2) equals

$$\frac{1}{p!}P_{i_1 i_2 \ldots i_p}P^{i_1 i_2 \ldots i_p},$$

where $P_{i_1 i_2 \ldots i_p}$ denotes the covariant components (i.e., the determinants constructed from the covariant components of the p vectors). From the results of Section 14,

$$V^2 = \begin{vmatrix} \mathbf{x}^2 & \mathbf{x}.\mathbf{y} \ldots \mathbf{x}.\mathbf{z} \\ \mathbf{y}.\mathbf{x} & \mathbf{y}^2 \ldots \mathbf{y}.\mathbf{z} \\ \cdots & \cdots \\ \mathbf{z}.\mathbf{x} & \mathbf{z}.\mathbf{y} \ldots \mathbf{z}^2 \end{vmatrix} = \begin{vmatrix} x^i x_i & x^j y_j \ldots x^k z_k \\ y^i x_i & y^j y_j \ldots y^k z_k \\ \cdots & \cdots \\ z^i x_i & z^j y_j \ldots z^k z_k \end{vmatrix} = x_i y_j \ldots z_k \begin{vmatrix} x^i x^j \ldots x^k \\ y^i y^j \ldots y^k \\ \cdots \\ z^i z^j \ldots z^k \end{vmatrix}$$

$$= \frac{1}{p!} P_{ij\ldots k} P^{ij\ldots k}$$

where the sum extends over all permutations of the p of the n indices $1, 2, \ldots, n$.

Sets of n vectors can be regarded as n-vectors; they have only one contravariant component and one covariant component. Each component merely changes its sign under a reversal.

16. Isotropic multivectors

A p-vector is said to be isotropic if its volume is zero but not all its components are zero; and if it spans a linear manifold of dimension not less than p. For a p-vector to be isotropic it is necessary and sufficient for there to be a vector in its p-dimensional linear manifold which is orthogonal to the manifold; such a vector must be orthogonal to itself, i.e., be isotropic. Thus if the volume of a p-vector is zero, there are p constants $\alpha, \beta, \ldots, \gamma$ not all zero which satisfy the equations

$$\begin{aligned} \alpha\mathbf{x}^2 + \beta\mathbf{y}.\mathbf{x} + \cdots + \gamma\mathbf{z}.\mathbf{x} &= 0 \\ \alpha\mathbf{x}.\mathbf{y} + \beta\mathbf{y}^2 + \cdots + \gamma\mathbf{z}.\mathbf{y} &= 0 \\ \cdots\cdots\cdots\cdots\cdots \\ \alpha\mathbf{x}.\mathbf{z} + \beta\mathbf{y}.\mathbf{z} + \cdots + \gamma\mathbf{z}^2 &= 0. \end{aligned}$$

This is the same as saying that there exists a non-zero vector $\alpha\mathbf{x} + \beta\mathbf{y} + \cdots + \gamma\mathbf{z}$ orthogonal to each of the vectors $\mathbf{x}, \mathbf{y}, \ldots, \mathbf{z}$. The converse holds. Another interpretation is to say that the manifold of the p-vector is tangential to the isotropic hypercone (formed of isotropic directions) along the line in which the vector $\alpha\mathbf{x} + \beta\mathbf{y} + \cdots + \gamma\mathbf{z}$ lies.

17. Supplementary multivectors

Given a non-isotropic p-vector, we say that an $(n - p)$-vector is *supplementary* to this p-vector if (a) its $(n - p)$-dimensional linear manifold consists of vectors orthogonal to the manifold of the p-vector, (b) the volume of the $(n - p)$-vector equals that of the p-vector, and (c) the volume of the n-hedron formed from the p vectors of the p-vector and the $n - p$ vectors of the $(n - p)$-vector is positive. Note that the $(n - p)$-dimensional manifold is uniquely determined by the p-vector; it can have no vector in common with the manifold of the p-vector, otherwise the latter would be isotropic.

Assume that $P^{12\ldots p} \neq 0$. The equations which express the fact that a vector $\mathbf{t}$ is orthogonal to p vectors $\mathbf{x}, \mathbf{y}, \ldots \mathbf{z}$ are

$$t_i x^i = 0, \qquad t_i y^i = 0, \qquad \ldots, \qquad t_i z^i = 0;$$

On eliminating $t_2, t_3, \ldots, t_{p+1}$, it follows that

$$t_1 P^{12\ldots p} + t_{p+1} P^{(p+1)2\ldots p} + \cdots + t_n P^{n23\ldots p} = 0.$$

It also follows that

$$t_2\mathrm{P}^{213\ldots p} + t_{p+1}\mathrm{P}^{(p+1)13\ldots p} + \cdots + t_n\mathrm{P}^{n13\ldots p} = 0,$$

$$t_p\mathrm{P}^{p12\ldots(p-1)} + t_{p+1}\mathrm{P}^{(p+1)12\ldots(p-1)} + \cdots + t_n\mathrm{P}^{n12\ldots(p-1)} = 0.$$

These are the equations satisfied by vectors in the manifold of the supplementary $(n-p)$-vector. If $\mathrm{Q}_{i_1 i_2 \ldots i_{n-p}}$ denote the covariant components of the $(n-p)$-vector, then since $\mathbf{t}$ lies in its manifold,

$$t_{i_1}\mathrm{Q}_{i_2 i_3 \ldots i_{n-p+1}} - t_{i_2}\mathrm{Q}_{i_1 i_3 \ldots i_{n-p+1}} + \cdots + (-1)^{n-p} t_{i_{n-p+1}}\mathrm{Q}_{i_1 i_2 \ldots i_{n-p}} = 0$$

where the set of suffixes $i_1, i_2, \ldots, i_{n-p+1}$ are successively put equal to the combinations $(1, p+1, p+2, \ldots, n)$, $(2, p+1, p+2, \ldots, n)$ etc. On comparing these equations with the preceding ones, it is seen that the $\mathrm{P}^{i_1 i_2 \ldots i_p}$ and the $\mathrm{Q}_{i_{p+1} i_{p+2} \ldots i_n}$ are proportional, when the permutation $(i_1 i_2 \ldots i_n)$ is even.

Thus we can write

$$\mathrm{P}^{i_1 i_2 \ldots i_p} = \lambda \mathrm{Q}_{i_{p+1} i_{p+2} \ldots i_n},$$

and also

$$\mathrm{P}_{i_1 i_2 \ldots i_p} = \mu \mathrm{Q}^{i_{p+1} i_{p+2} \ldots i_n};$$

but the p-vector and its supplementary $(n-p)$-vector have the same magnitude, which requires $\lambda\mu = 1$; on the other hand, using the Laplace expansion for a determinant, the n-vector formed from vectors of the p-vector and of the $(n-p)$-vector has the measure

$$\mathrm{V} = \frac{1}{\sqrt{g}} \sum \pm \mathrm{P}_{i_1 i_2 \ldots i_p} \mathrm{Q}_{i_{p+1} \ldots i_n} = \frac{1}{\lambda\sqrt{g}} \sum \mathrm{P}_{i_1 i_2 \ldots i_p} \mathrm{P}^{i_1 i_2 \ldots i_p};$$

also

$$\mathrm{V} = \sqrt{g} \sum \pm \mathrm{P}^{i_1 i_2 \ldots i_p} \mathrm{Q}^{i_{p+1} \ldots i_n} = \frac{\sqrt{g}}{\mu} \sum \mathrm{P}^{i_1 i_2 \ldots i_p} \mathrm{P}_{i_1 i_2 \ldots i_p}$$

where the sums on the right-hand side extend over all combinations $i_1, i_2, \ldots, i_p$ of the indices taken p at a time. Since V and this latter sum are both positive, $\lambda = 1/\sqrt{g}$ and $\mu = \sqrt{g}$, from which it follows that

$$\left.\begin{aligned} \mathrm{P}^{i_1 i_2 \ldots i_p} &= \frac{1}{\sqrt{g}} \mathrm{Q}_{i_{p+1} i_{p+2} \ldots i_n} \\ \mathrm{P}_{i_1 i_2 \ldots i_p} &= \sqrt{g}\, \mathrm{Q}^{i_{p+1} i_{p+2} \ldots i_n} \end{aligned}\right\} \tag{10}$$

NOTE. In defining the $(n-p)$-vector supplementary to a given p-vector we have assumed that the latter was not isotropic, but the above formulae allow the definition to be extended to all cases. It is then possible for a p-vector to equal its supplementary vector (where of course $n = 2p$).

18. Sum of p-vectors

Let us consider a set of p-vectors; it is convenient to take two such sets as being equal if the sums of the components with the same indices of all the p-vectors in each set are the same. We shall use the same notation for the sums as for p-vectors and refer to these sums as "components" of the set. It will be convenient to say that these ${}^nC_p$ components define a p-vector; the p-vectors originally defined will now be called "simple p-vectors". The supplementary vector of a non-simple p-vector will be defined by the same formulae as for a simple p-vector. Analytically a general p-vector can be defined as an antisymmetric set of ${}^nC_p$ numbers $P^{i_1 i_2 \ldots i_p}$, each labelled by p distinct indices, i.e., under a permutation of the indices a component is unaltered or merely changed in sign, depending on whether the permutation is even or odd.

IV. BIVECTORS AND INFINITESIMAL ROTATIONS

A bivector is defined by $[n(n-1)]/2$ quantities $a^{ij} = -a^{ji}$; the necessary and sufficient condition for a bivector to be simple is that its components satisfy

$$a^{ij}a^{kh} + a^{jk}a^{ih} + a^{ki}a^{jh} = 0 \qquad (i, j, k, h = 1, 2, \ldots, n).$$

19. Infinitesimal rotations

Bivectors occur when the family of rotations depending on a parameter t is considered. Suppose that the space is referred to a Cartesian frame of reference $(\mathbf{e}_1, \mathbf{e}_2, \ldots, \mathbf{e}_n)$, then the velocity at an instant t of a point x of the space has as its components expressions linear in the co-ordinates of the point. Thus let

$$(x^i)^1 = \alpha^i_k(t)x^k$$

be the equation giving the co-ordinates $(x^i)^1$ at time t of a point which had co-ordinates x^k at the initial time t_0. Assume that the functions $\alpha^i_k(t)$ have first derivatives; then the velocity at any time t of a point which was initially (x) has as its components

$$v^i = \frac{\mathrm{d}}{\mathrm{d}t}(\alpha^k_i(t)) \,.\, x^k;$$

and since the x^k are linear combinations of the co-ordinates $(x^i)^1$ at time t, the v^i must also be linear combinations of these latter co-ordinates.

With a slight change of notation, suppose that the velocity at time t when the co-ordinates are x^i is given by $v^i - a^i_k x^k$. Then, since the velocity must be perpendicular to the vector from the origin 0 to the point x,

$$x_i v^i \equiv a^i_k x_i x^k \equiv a_{ik} x^i x^k = 0.$$

This implies that $a_{ij} + a_{ji} = 0$. The a^i_k are the *mixed* components of a bivector.

An "infinitesimal rotation" is a variable rotation which can be considered as taking place in an infinitely short time-interval from t to $t + \mathrm{d}t$, during which each point x undergoes an elementary displacement $\delta\mathbf{x} = \mathbf{v}\,dt$ where $\mathbf{v}$ is the velocity at time t.

Any infinitesimal rotation can thus be defined by

$$\delta x^i = a^i_k x^k \tag{11}$$

where the a^i_k are the mixed components of a bivector. Infinitesimal rotations are linearly dependent on $[n(n-1)]/2$ parameters.

CHAPTER II

TENSORS; LINEAR REPRESENTATIONS OF GROUPS; MATRICES

I. DEFINITION OF TENSORS

20. First example of a linear representation

Given a Euclidean space E_n referred to a fixed Cartesian frame of reference $(\mathbf{e}_1, \mathbf{e}_2, \ldots, \mathbf{e}_n)$, we have considered the linear transformations S which represent the effect produced on a vector $\mathbf{x}$ by a given rotation R. These linear transformations possess the obvious property that, if the transformations S and S′ correspond to the rotations R and R′, the transformation which corresponds to the rotation R′R, obtained by carrying out successively first R then R′, is the transformation S′S obtained by carrying out first the transformation S, then the transformation S′. The set of transformations S constitutes what we shall call a linear representation of the group of rotations.

If we change the frame of reference to a new frame of reference not resulting from the old one by a rotation or a reversal, the rotation R which was represented by a linear transformation S would be represented by another distinct linear transformation T; the set of transformations T would constitute a new linear representation of the group of rotations. It is, however, clear that there is a strict relationship between these two representations, which denote analytically the same geometrical operations carried out on the same geometrical objects. The relationship is as follows: if x^i are the components of a vector referred to the first frame of reference, and y^i the components of this vector referred to the second frame of reference, we can pass from the x^i to the y^i by a fixed linear transformation σ; each transformation T is deduced from the corresponding transformation S by applying to the variables x^i and their transforms $(x^i)^1$ the same fixed linear transformation, viz. that which allows us to pass from the x^i to the y^i. To be more precise, we pass

from the y^i to the x^i by the transformation σ^{-1},
from the x^i to the $(x^i)^1$ by the transformation S,
from the $(x^i)^1$ to the $(y^i)^1$ by the transformation σ.

It follows that the passage from the y^i to the $(y^i)^1$ is carried out by the successive transformations σ^{-1}, S, σ, which is expressed by the formula

$$T = \sigma S \sigma^{-1}.$$

We shall say that the linear representation T of the group of rotations is *equivalent* to the linear representation S, and the equivalence is shown by the existence of a fixed linear transformation σ such that to each transformation S of the first representation there corresponds the transformation $\sigma S \sigma^{-1}$ of the second.

21. General definition of linear representations of the group of rotations

Let two vectors (x^i) and (y^j) be referred to the same Cartesian frame of reference and let us consider the n^2 products $x^i y^j$; as a result of a rotation they obviously undergo a linear transformation Σ, which also possesses the property that if Σ and Σ' correspond to the rotations R and R′, the transformation $\Sigma'\Sigma$ corresponds to R′R. The n^2 quantities $x^i y^j$ therefore provide a new linear representation of the group of rotations, completely distinct from the two previous ones.

In general, a family of linear transformations S applied to r variables $u_1, u_2, \ldots, u_r$ will provide a linear representation of the group of rotations if to each rotation R there corresponds a transformation S determined in such a way that to the product R′R of two rotations there corresponds the product S′S of the two corresponding transformations. The integer r is called the *degree* of the representation.

22. Equivalent representations

Two linear representations of the group of rotations are *equivalent*:

(1) if they have the same degree;

(2) if we can pass from the first to the second by applying to the variables of the first a non-degenerate fixed linear transformation σ (i.e., with non-zero determinant).

If S and T correspond to R in the two representations, we then have

$$T = \sigma S \sigma^{-1}$$

where σ denotes a fixed transformation.

A linear representation is said to be *faithful* if two distinct linear transformations correspond to two distinct rotations.

23. The notion of a Euclidean tensor

It may be asked whether the variables $u_1, u_2, \ldots, u_r$ which occur in a linear representation of the group of rotations are capable of a concrete interpretation. We may assert that a set of r numbers $(u_1, u_2, \ldots, u_r)$ constitutes an object, and agree that the effect produced on the object $(u_1, u_2, \ldots, u_r)$ by the rotation R is to bring it into coincidence with the object $(u'_1, u'_2, \ldots, u'_r)$, whose *components* u'_i are deduced from the components u_i by the transformation S which corresponds to R. This convention is *consistent*, because the effect of the rotation R′ on the object (u') is the same as that of the rotation R′R on the initial object (u). Furthermore we can restrict the family of objects (u) under consideration by making the components $u_1, u_2, \ldots, u_r$ satisfy certain fixed algebraic relationships, subject to the double condition that:

(i) the components $u_1, u_2, \ldots, u_r$ of the objects of the restricted family do not satisfy any linear relationship with constant coefficients;

(ii) the algebraic relationships which determine the restricted family should remain invariant under the transformations S of the linear representation.

The family of objects thus specified will be called a *Euclidean tensor* which we shall consider as being associated with the point O. Two Euclidean tensors will be said to be equivalent if they arise from the same linear representation of the group of rotations or from two equivalent linear representations.

For example, the set of all vectors drawn from the origin, the set of the vectors of length 1, and the set of isotropic vectors, constitute three equivalent tensors; if the vectors of each of these sets are represented by components referred to a Cartesian frame of reference, then the second set is characterised by the relationship $\Phi(x) = 1$, and the third by the relationship $\Phi(x) = 0$, where $\Phi(x)$ denotes the fundamental form.

It is important to observe that the tensor is defined not by the nature of the objects which compose it, but by the choice of components which define it analytically. For example, a pair of *real* opposed vectors $\mathbf{x}$ and $-\mathbf{x}$ may be represented analytically either by the $n(n + 1)/2$ monomials $x^i x^j$, or by the $n(n + 1)(n + 2)(n + 3)/24$ monomials $x^i x^j x^k x^l$; each of these analytical representations defines a Euclidean tensor, but these two tensors are not equivalent.

Naturally what has just been said of the group of rotations could be said of the group of rotations and reversals, but it is important to observe that a tensor for the group of rotations is not necessarily a tensor for the group of rotations and reversals; we shall soon see an example of this (Section 51). There are therefore grounds for distinguishing Euclidean tensors in the strict sense, which provide linear representations of the group of rotations, from Euclidean tensors in the more general sense, which provide linear representations of the group of rotations and reversals. In a real pseudo-Euclidean space, this classification could be developed further, according to whether the group of proper rotations, the group of all the rotations, or the group of proper rotations and proper reversals, etc., is under consideration.

24. Another point of view

We can consider tensors from a slightly different point of view which is that usually taken in tensor calculus. Let us take again the equations that define the effect of a rotation R on a vector $\mathbf{x}$ which is referred to a given Cartesian frame of reference R. The equations of the transformation S which represent the rotation R may be interpreted as providing the passage from the components x^i of the vector to the components $(x^i)'$ of the *same vector*, but referred to the frame of reference R′ which is obtained from R by applying the rotation R. In this approach then, the transformation S is taken as an operation changing the co-ordinates, the substitution S being that which allows us to pass from the old frame of reference to the new one, *which is expressed analytically when it is referred to the old frame of reference.* It is of course important to observe that the old and the new frames of reference are the same from the point of view of the group of rotations. The linear representation of the group of rotations therefore expresses the changes of co-ordinates operating on the components of an arbitrarily chosen vector, but the *different systems of co-ordinates under consideration, not being systems of arbitrarily chosen Cartesian co-ordinates*, must all be equivalent with reference to the group of rotations, the corresponding frames of reference being directly equal.

We can consider transformations of the components of a vector referred to Cartesian frames of reference connected by relations with absolutely arbitrary coefficients. The linear transformations thus obtained do not provide a linear representation of the group of rotations, but of a wider group, that of *affine transformations*; they correspond to absolutely arbitrary linear transformations. From this point of view vectors and p-vectors are *affine tensors*, in the sense that they provide linear representations of the group of affine transformations (full linear group). This is furthermore the way in which they are regarded in classical tensor calculus.

II. TENSOR ALGEBRA

The notion of a tensor may be generalised to any group G; a tensor relative to G may be defined by a linear representation of G, but the entities represented by this tensor may be capable of several different concrete interpretations.

Whatever group G is under consideration, tensor calculus comprises certain simple operations and satisfies certain general theorems which we shall briefly indicate.

25. Addition of two equivalent tensors

Given two equivalent tensors with components $x^1, x^2, \ldots, x^r$ and $y^1, y^2, \ldots, y^r$, these components being chosen so that to any operation of G there corresponds the same linear transformation of the variables x^i and the variables y^i, the tensor having as its components $x^i + y^i$ is called the sum of the two tensors; this sum constitutes a tensor equivalent to the given tensors. More generally, the quantities $mx^i + ny^i$, where m and n are two fixed constants, define a tensor equivalent to the given tensors.

26. Multiplication of any two tensors

Given two tensors, equivalent or not, with components $x^1, x^2, \ldots, x^r$ and $y^1, y^2, \ldots, y^s$, the tensor with components $x^i y^j$ is called the *product* of these tensors.

27. Some fundamental theorems

THEOREM I. *Let $x^1, x^2, \ldots, x^r$ be the components of a tensor, and $y_1, y_2, \ldots, y_r$ variables which are transformed by the operations of the group G in such a way as to leave the sum $x^i y_i$ invariant; then the quantities y_i define a tensor whose nature depends only on the nature of the first tensor.**

In fact let

$$(x^i)' = a^i_k x^k$$

be the linear transformations of the components x^i under an operation of G; the transformations $y_i \to y'_i$ of the quantities y_i satisfy the identity

$$a^i_k x^k y'_i = x^k y_k,$$

hence

$$y_k = a^i_k y'_i;$$

it follows that the y'_i are obtained from the y_k by a linear transformation, which depends only on the linear transformation carried out on the x_i.

THEOREM II. *Let $x^1, x^2, \ldots, x^r$ be the components of a tensor and $y^1, y^2, \ldots, y^s$ quantities which are transformed by the operations of the group G so that the h expressions*

$$z_\alpha \equiv c^k_{i\alpha} x^i y_k$$

are transformed like the components of a tensor: under these conditions the hr quantities $c^k_{i\alpha} y_k (i = 1, 2, \ldots, r;\ \alpha = 1, 2, \ldots, h)$ form a tensor whose nature depends only on the nature of the tensor (x) and of the tensor (z). The degree of this tensor is equal to the number of the independent linear forms $c^k_{i\alpha} y_k$.

Let us in fact suppose that, by an operation of G, the x^i and the z_α undergo the linear transformations

$$(x^i)' = a^i_j x^j;$$

$$z'_\alpha = b^\lambda_\alpha z_\lambda:$$

we will then have

$$c^k_{i\alpha} a^i_j x^j y'_k = b^\lambda_\alpha c^k_{i\lambda} x^i y_k,$$

whence

$$c^k_{i\alpha} a^i_j y'_k = b^\lambda_\alpha c^k_{j\lambda} y_k \; (\alpha = 1, 2, \ldots, h;\ j = 1, 2, \ldots, r).$$

---

* We say that two tensors are of the same nature if they are equivalent.

These equations can be solved for the quantities $c^k_{i\alpha}y'_k$ which are then linearly expressed in terms of the $c^k_{j\lambda}y_k$.

28. Applications

We have seen (Section 19) that every infinitesimal rotation depends on $n(n-1)/2$ quantities $a_{ij} = -a_{ji}$, the velocity v_i of the point x^i being the vector

$$v_i = a_{ik}x^k.$$

We pointed out that the quantities a_{ij} can be regarded as the components of a bivector. In reality we had not proved this; we merely assured ourselves that under a rotation applied simultaneously to the vector **x** and to its velocity **v** the a_{ij} transform in the same way as the components of a bivector. Now let us consider an arbitrary vector **y** and the scalar product $\mathbf{y}\,.\,\mathbf{v} = a_{ik}y^ix^k$. This sum is an invariant; also, the quantities y^ix^k constitute a tensor, therefore the same is true of the a_{ik} (Theorem I) and the law of transformation of the a_{ik} depends only on the law of transformation of the y^ix^k. Furthermore, let **u** and **v** be two arbitrary vectors; then the sum

$$(u_iv_k - v_iu_k)y^ix^k = u_iy^iv_kx^k - v_iy^iu_kx^k$$

is an invariant; therefore the quantities $u_i v_k - v_iu_k$ transform in the same way as the a_{ik} and they constitute the components of a simple bivector: *therefore an infinitesimal rotation defines a tensor which is equivalent to a bivector.*

III. REDUCIBLE AND IRREDUCIBLE TENSORS

29. Definitions

A tensor with respect to a group G is said to be *reducible* if it has r components $u_1, u_2, \ldots, u_r$ and if it is possible to form $\rho < r$ linear combinations with *complex* constant coefficients such that these combinations themselves have the characteristics of a tensor, i.e., they are transformed linearly amongst themselves by every operation of the group G.

All tensors that are not reducible are said to be irreducible.

A tensor is said to be *completely reducible* if it is possible by means of a suitable linear change of components to divide these into a certain number of sets

$$\begin{gathered} x_1, x_2, \ldots, x_p; \\ y_1, y_2, \ldots y_q; \\ z_1, z_2, \ldots, z_r; \\ \ldots\ldots\ldots\ldots\ldots, \end{gathered}$$

in such a manner that the components in each set transform amongst themselves, and transform in an irreducible way.

It is clear that if a tensor is irreducible, all tensors equivalent to it are also irreducible; and the same holds for complete reducibility.

A tensor with respect to a group G which provides a linear representation of G gives a linear representation of any subgroup g of G; it is thus also a tensor with respect to g; if it is irreducible with respect to g it is obviously also irreducible with respect to G, but the converse is not always true. We have already quoted an example where G is the group of rotations and reversals, and g the group of rotations.

30. A criterion for irreducibility

The concept of irreducibility can be introduced in a different way. We consider the r components of a tensor as components of a vector in a space of dimension r. If the tensor is reducible, then, there exist $\rho < r$ linear combinations of these r quantities u_i which transform linearly amongst themselves under the operations of G. This amounts to saying that the linear manifold π which contains the origin consisting of those vectors which make these ρ linear forms zero is invariant under the operations of G. Conversely if G leaves invariant a linear manifold π which contains the origin, then the left-hand sides of the quations which define π are transformed linearly amongst themselves under the substitutions G and thus the tensor is reducible. *In order for a tensor to be irreducible, it is necessary and sufficient that the substitutions should not leave invariant any linear manifold which contains the origin.*

31. A property of irreducible tensors

It is possible to deduce from this a property of irreducible tensors which will be useful to us later. We again represent each particular member of the set of tensors we are considering by a vector $\mathbf{u}$ in a space E_r of dimension r. We do not thus necessarily obtain all vectors in E_r, but as there is no linear homogeneous relationship between the components $u_1, u_2, \ldots, u_r$ of the tensors in this set, any vector in E_r can be obtained as a linear combination of vectors which represent tensors in this set. We now apply to the vector $\mathbf{u}_0$, which represents a particular tensor in the set, all the operations of the group G. We thus obtain a certain family of vectors. These vectors cannot all belong to the same linear manifold π which contains the origin. If in fact we suppose that π is the smallest of these linear manifolds, every vector $\mathbf{u}$ of the manifold π is a linear combination of a certain number of vectors in the family, for example, those vectors which are obtained from $\mathbf{u}_0$ by applying the substitutions $S_1, S_2, \ldots, S_p$. Then if we apply to this vector $\mathbf{u}$ that operation of G which is represented by the substitution S, we obtain the same linear combination with vectors $SS_1\mathbf{u}_0, SS_2\mathbf{u}_0, \ldots, SS_p\mathbf{u}_0$ all by hypothesis in π. It follows that all substitutions S leave π invariant; since the tensor is irreducible, this is not possible unless the manifold π comprises the whole space.

32. Problem

Given an irreducible tensor with r components $u_1, u_2, \ldots, u_r$, is it possible to make a linear change of the components (i.e., a change of basis) so that the new components transform under the same linear substitutions S as do the original components for all elements of the group G? Let σ be the linear substitution giving the new components in terms of the old, then for all substitutions the equations

$$\sigma S \sigma^{-1} = S \quad \text{or} \quad \sigma S = S\sigma \tag{1}$$

must hold.

In the space E_r consisting of vectors $\mathbf{u}$ there always exists a vector $\mathbf{u}_0$ which as a result of the substitution σ is merely multiplied by a constant, m. Let the equations

$$v_i = c_i^k u_k$$

define σ; then we have to find r numbers u_i which satisfy

$$c_i^k u_k = m u_i \qquad (i = 1, 2, \ldots, r).$$

This is possible, subject to the single condition that the determinant

$$\begin{vmatrix} c_1^1 - m & c_1^2 & \ldots & c_1^2 \\ c_2^1 & c_2^2 - m & \ldots & c_2^r \\ \cdots & \cdots & \cdots & \cdots \\ c_r^1 & c_r^2 & \ldots & c_r - m \end{vmatrix}$$

is zero; we can take any root of this polynomial to be m. This gives the result

$$\sigma \mathbf{u}_0 = m \mathbf{u}_0$$

and equation (1) implies that

$$\sigma - (S\mathbf{u}_0) = S(\sigma \mathbf{u}_0) = m(S\mathbf{u}_0);$$

the operation σ thus merely multiplies $S\mathbf{u}_0$ by m. But since the tensor is irreducible, any vector in E_r can be written as a linear combination of the vectors $S\mathbf{u}_0$ and thus the effect of σ will merely be to multiply it by m. Thus $v_i = m u_i$.

THEOREM. *The only linear change of variables which leaves invariant all the substitutions of an irreducible linear representation of a group* G *consists of a multiplication of each of the variables by the same constant factor m.*

33. Discussion of the irreducible tensors contained in a completely reducible tensor

We apply the above to the solution of the following problem. Let a tensor be completely reducible with respect to a group G. Suppose, for example, that by a suitable choice of components it decomposes into five irreducible tensors of which the first three are equivalent to each other, and the last two are also equivalent. We denote the components of these tensors by

$$x_1, x_2, x_3, \ldots, x_p,$$

$$y_1, y_2, y_3, \ldots, y_p,$$

$$z_1, z_2, z_3, \ldots, z_p,$$

$$u_1, u_2, u_3, \ldots, u_q,$$

$$v_1, v_2, v_3, \ldots, v_q.$$

We can also suppose that the components x_i, y_i and z_i are chosen in such a way that the linear substitutions undergone by the x_i, y_i and z_i are identical, and similarly for the u_i and the v_i.

With these assumptions let us look for all the irreducible tensors which can be obtained from the given completely reducible tensor. The components of such an irreducible tensor will be linear combinations of the x_i, y_i, z_j, u_i, v_i. Let

$$\sum a_i x_i + \sum b_j y_j + \sum c_k z_k + \sum h_\alpha u_\alpha + \sum k_\beta v_\beta$$

be one of these components. If one of the coefficients a_i is different from zero, it is impossible for all the components x_i to be missing from any other of the linear combinations which form the components of the tensor we are looking for; otherwise the set of all these components in which none of the letters x_i occur would itself have the characteristics of a tensor. On the other hand the different quantities $\Sigma a_i x_i$ transform linearly amongst themselves, and since the x_i form an irreducible tensor, it follows that the tensor we are looking for contains exactly p components which can be put in the form

$$x_1 + \cdots,$$
$$x_2 + \cdots,$$
$$\cdots\cdots\cdots$$
$$x_p + \cdots,$$

where the dots represent linear combinations of the y_i, z_i and u_i, v_i. The p linear combinations of the y_i which occur here (assuming that they are not all zero) must transform in the same way as $x_1, x_2, \ldots, x_p$ and it follows (Section 32) that they must necessarily be of the form $my_1, my_2, \ldots, my_p$. A similar result holds for the linear combinations of the z_i. As for the variables u_i, they cannot occur, since the p linear combinations would have to transform amongst themselves like the x_i; this would require that $q = p$ and in addition that the u_i form a tensor equivalent to the x_i which is contrary to hypothesis. We conclude that the only irreducible tensors which can be obtained from the given tensor are of the form $lx_i + my_i + nz_i$ and $\alpha u_i + \beta v_i$ where l, m, α, β are constants.

34. Application

An important result can be deduced from the above discussion. Suppose we have a class of objects which are transformed amongst themselves by the operations of the group G, and that we can associate with each of the objects in the class a certain number, r, of quantities u_α such that:

(i) Under all operations of G, the u_α undergo a definite linear substitution S.

(ii) These linear substitutions S form a linear representation of G.

Can there be one or more linear relations with constant coefficients amongst the u_α?

The postulates can be expressed by saying that in a certain r-dimensional space E_r the r components x_α of certain vectors $\mathbf{x}$ define a tensor $\mathscr{T}$. The vectors $\mathbf{u}$ associated with the objects in the class only form a part of the space. But since the substitutions S transform them amongst themselves, it follows that the smallest subspace containing all the vectors $\mathbf{u}$ is invariant under the substitutions S. The left-hand sides of the equations which define this subspace thus form a tensor contained in $\mathscr{T}$. It follows that *any linear relations which exist amongst the u_α can be obtained by equating to zero all the components of a certain tensor contained in* $\mathscr{T}$.

Suppose for example that the tensor $\mathscr{T}$ can be decomposed into three irreducible tensors which are equivalent to each other, and suppose that the components

$$x_1, x_2, \ldots, x_\rho,$$

$$y_1, y_2, \ldots, y_\rho,$$

$$z_1, z_2, \ldots, z_\rho,$$

of these three tensors have been chosen in such a manner that the x_i, y_i, and z_i all undergo the same substitution for each operation in G. Relations of the type we are seeking can be obtained by writing one or more sets of ρ relations of the form

$$ax_i + by_i + cz_i = 0 \qquad (i = 1, 2, \ldots, \rho)$$

where a, b, and c are constant coefficients.

35. A noteworthy theorem

We shall apply the preceding result to prove a well-known theorem.

Burnside's Theorem*. *There is no linear relation with constant coefficients connecting the various substitutions* S *which form an irreducible linear representation of a group* G.

Let S of degree n, with elements u_i^j, denote the general substitution of the representation; then

$$x_i' = u_i^j x_j.$$

Let A be a particular substitution in the representation, then $S' = AS$ will also occur and its coefficients $(u_i^j)'$ can be obtained from u_i^j by the equations

$$(u_i^j)' = a_i^k u_k^j. \tag{2}$$

To each operation a of G there corresponds a linear substitution (2) of the n^2 quantities u_i^j; these linear substitutions provide a representation of the group G, since to successive operations a and b of G correspond the successive operations

$$S' = AS, \qquad S'' = BS' = (BA)S.$$

* *London Math. Soc. Proc.* **3**, 1905, p. 430.

We now apply the theorem of the preceding section. If we give j a fixed value in the equation (2), we see that the quantities $u_1^j, u_2^j, \ldots, u_n^j$ are transformed by the substitutions in the same manner as components of the original tensor x_k. But since this latter tensor is irreducible we deduce that the tensor u_i^j decomposes into n irreducible tensors all equivalent to each other. It follows that all possible linear relations between the u_i^j are obtained by writing one or more systems of n equations of the form

$$m_1 u_i^1 + m_2 u_i^2 + \cdots + m_n u_i^n = 0 \qquad (i = 1, \ldots, n),$$

where the m_k are constants. But such relations are impossible since they would make the determinant of the substitution zero. The theorem is thus proved.

Let us take as an example the effect of rotations on vectors in a real plane (x, y):

$$x' = x \cos \alpha - y \sin \alpha,$$

$$y' = x \sin \alpha + y \cos \alpha.$$

As the coefficients of the substitutions are not independent, the vector does not provide an irreducible tensor. In fact it can be decomposed into two *semi-vectors* $x + iy$ and $x - iy$ each with a single component.

36. A criterion of irreducibility

We can deduce from Burnside's theorem a criterion for the irreducibility of a linear representation.

THEOREM. *For a linear representation of any group whatsoever to be irreducible, it is necessary and sufficient that there should be no linear relation, with coefficients not all zero, between the different substitutions which comprise the given representation.*

By Burnside's theorem the condition is necessary. It is obviously sufficient, since if it is fulfilled it is not possible to extract from the given tensor $\mathscr{T}$ of degree r, any tensor $\mathscr{T}'$ of lesser degree ρ, because on taking the components of $\mathscr{T}'$ as the first ρ basis components of $\mathscr{T}$, the coefficients of the other $r - \rho$ components in each of the first ρ equations which define the substitutions S would all be zero.

IV. MATRICES

37. Definition

Every linear transformation S of the variables x_i is of the form

$$x'_i = a_i^k x_k \qquad (i = 1, 2, \ldots, n)$$

and can be represented by a *matrix*, i.e., the square diagram with n rows and n

columns of the coefficients of the transformation. We shall denote this matrix by the symbol S. If a new transformation S′ is carried out on the transformed variables to give

$$x''_i = b^k_i x'_k,$$

the resulting transformation S″ = S′S is defined by the formulae

$$x''_i = b^k_i a^h_k x_h.$$

From this we deduce a *law of multiplication* for matrices, the elements c^j_i of the matrix S″S being

$$c^j_i = b^k_i a^j_k.$$

More generally, we may consider rectangular matrices and define the product S′S by the preceding formulae, *on condition that the number of columns of* S′ *is equal to the number of rows of* S. In particular if x represents the matrix with n rows and 1 column whose elements are $x_1, x_2, \ldots, x_n$ and x^1 is the analogous matrix resulting from a transformation S, we have

$$x' = \mathrm{S}x.$$

38. Addition, multiplication by *m*

The sum of two matrices S and S′, which have the same numbers of rows and columns, is the matrix whose elements are obtained by adding the elements of S and S′ which occupy the same position; this matrix is denoted by S + S′.

The product of a matrix S and of a number m, which we shall denote by mS, is the matrix obtained by multiplying each of the elements of S by m.

39. Remark on the calculation of a product of matrices

It is important to make here an observation concerning the calculation of the product S′S of two linear transformations. We might expect that, in order to obtain the variable x''_i resulting from the transformation S′S, we could first calculate the variable $x'_i = a^k_i x_k$ resulting from the transformation S, then, in the linear form in the $x_1, x_2, \ldots, x_k$ so obtained, replace each x_k by its transform by S′, i.e., by $b^h_k x_h$. But the result *would be wrong*; we would in fact have

$$x''_i = a^k_i b^h_k x_h,$$

instead of

$$x''_i = b^k_i a^h_k x_h.$$

It is easily seen that if the above procedure is to be employed, it is important to apply the operations S′S in the reverse order to that indicated, carrying out first the operation S′, then the operation S. An analogous result holds for the product of any number of transformations.

40. Transposed matrices, inverse matrices

The matrix which is obtained from a matrix S by the exchange of rows and columns is called the *transposed matrix* (*transpose*) of S and denoted by S^{T}.

The matrix x^T is thus the matrix with one row and n columns consisting of elements $x_1, x_2, \ldots, x_n$.

When each matrix is replaced by its transpose, the equation

$$S'' = S'S$$

must be modified to

$$S''^T = S^T S'^T;$$

in particular, the formula $x' = Sx$ may be written as $x'^T = x^T S^T$.

The matrix representing the inverse linear transformation of the transformation represented by S is called the *inverse matrix* of the given square matrix S (whose determinant must not be zero), and denoted by S^{-1}. We have

$$SS^{-1} = S^{-1}S = 1,$$

where (without risk of ambiguity) we use 1 to denote the matrix all of whose elements are zero except those on the principal diagonal, which are equal to 1. For all S, we have

$$S1 = 1S = S.$$

A square matrix all of whose elements which are not on the principal diagonal are zero is called a diagonal matrix.

41. Equivalent matrices

We have seen (Sections 20, 22) that if, in a linear transformation S, the primitive variables and the transformed variables are made to undergo the same linear transformation A, the transformation ASA^{-1} is obtained. We shall say, in a general way, that the "transform" of the square matrix S by the square matrix A which has a non-zero determinant and is of the same order, is the matrix $S' = ASA^{-1}$; we say further that the matrices S and ASA^{-1} are *equivalent*.

Two equivalent square matrices have the same determinant. The transform of a product matrix S′S by the matrix A is the product of the transforms of S′ and S: this results from the equation

$$A(S'S)A^{-1} = (AS'A^{-1})(ASA^{-1}).$$

In the same way the inverse of the transform of S is the transform of the inverse of S; this results from the relation

$$(ASA^{-1})(AS^{-1}A^{-1}) = ASS^{-1}A^{-1} = AA^{-1} = 1.$$

42. Characteristic equation; Eigenvalues

The *characteristic equation* of a square matrix S is the equation in λ obtained by subtracting λ from all the elements on the principal diagonal of S, and equating to zero the determinant of the square matrix $S - \lambda 1$ so obtained. The roots of the characteristic equation are the *eigenvalues* of the matrix S.

Two equivalent matrices S and ASA^{-1} have the same characteristic equation. This results from the relations

$$ASA^{-1} - \lambda = A(S - \lambda)A^{-1},$$

whence

$$|ASA^{-1} - \lambda| = |A||S - \lambda||A^{-1}| = |S - \lambda|.$$

If S is the matrix of a linear transformation in a vector space E_n, then saying that λ_0 is an eigenvalue of S implies that there is a non-zero vector **x** whose transform by S is just λ_0 times the vector, i.e.,

$$S\mathbf{x} = \lambda_0 \mathbf{x}.$$

43. Unitary matrices

A square matrix U with complex elements is said to be *unitary* if the linear transformation which it represents leaves invariant the sum of the squares of the moduli of the variables, i.e., the product $\bar{x}^T x$ where $\bar{x}$ is the complex conjugate matrix of x with 1 column and n rows. Now we have

$$\bar{x}^{1T} x^1 = \bar{x}^T \bar{U}^T U x;$$

the condition for a square matrix to be unitary is therefore

$$\bar{U}^T U = 1 \quad \text{or} \quad \bar{U}^T = U^{-1}:$$

the transpose of the complex conjugate of U *is equal to the inverse of* U.

THEOREM. *Every unitary matrix can be transformed by a suitable unitary matrix into a diagonal matrix with all its elements of unit modulus.*

Before proving this theorem, let us agree to define the scalar product of vectors **x** and **y** by the quantity $\bar{x}^T y$. This scalar product changes into its complex conjugate when the order of vectors is changed: the scalar square of a vector is equal to the sum of squares of the moduli of its components: it can therefore only be zero if the vector is zero. Two vectors will be said to be orthogonal if their scalar product is zero. Let us remark finally that every unitary matrix U leaves invariant the scalar product of two vectors: we have in fact

$$\bar{x}^{1T} y^1 = \bar{x}^T \bar{U}^T U y = \bar{x}^T y.$$

Let λ be an eigenvalue of U: there exists a non-zero vector **x** which the matrix U reproduces multiplied by λ:

$$Ux = \lambda x;$$

by transposition and passage to the complex conjugate, we deduce from this that

$$\bar{x}^T \bar{U}^T = \overline{\lambda x}^T,$$

and hence by multiplication that

$$\bar{x}^T \bar{U}^T U x = \bar{x}^T x = \lambda \bar{\lambda} \bar{x}^T x$$

which gives $\bar{\lambda}\lambda = 1$. *All the eigenvalues of* U *have therefore modulus equal to* 1.

It should be noted moreover that *the transform of* U *by a unitary matrix* C *is still unitary:*

$$(\overline{CUC^{-1}})^T = \bar{C}^{-1T} \bar{U}^T \bar{C}^T = CU^{-1}C^{-1} = (CUC^{-1})^{-1}.$$

Let us now denote by $\mathbf{e}_1$ a vector, which may be taken to be a unit vector, which the matrix U reproduces multiplied by λ_1; the matrix U leaves invariant the subspace formed by vectors orthogonal to $\mathbf{e}_1$; there exists in this subspace at least one vector $\mathbf{e}_2$, which may be supposed to be a unit vector, which U reproduces multiplied by λ_2. The matrix U leaves invariant the subspace which is formed by vectors orthogonal to $\mathbf{e}_1$ and $\mathbf{e}_2$; in this subspace there exists at least one vector $\mathbf{e}_3$ which may be supposed to be a unit vector, which U reproduces multiplied by λ_3, and so on. Thus a new basis $\mathbf{e}_1, \mathbf{e}_2, \ldots, \mathbf{e}_n$ is arrived at. Each vector is of the form

$$y_1\mathbf{e}_1 + y_2\mathbf{e}_2 + \cdots + y_n\mathbf{e}_n,$$

and it is easily verified that the scalar square of this vector is $y_1\bar{y}_1 + y_2\bar{y}_2 + \cdots + y_n\bar{y}_n$; the transformation C which defines the passage from the x_i to the y_i is therefore unitary, and the matrix U′, which is the transform of U by C, operating on the y_i, transforms the vector $\Sigma y_i\mathbf{e}_i$ into $\Sigma\lambda_i y_i\mathbf{e}_i$: it is therefore diagonal, its elements being the λ_i, all having unit modulus.

44. Orthogonal matrices

A square matrix O is said to be *orthogonal* if the linear transformation which it represents leaves invariant the sum of the squares of the variables, i.e., the quantity $x^{\mathrm{T}}x$ is invariant. The equality

$$x^{\mathrm{T}}\mathrm{O}^{\mathrm{T}}\mathrm{O}x = x^{\mathrm{T}}x$$

gives

$$\mathrm{O}^{\mathrm{T}}\mathrm{O} = 1 \quad \text{or} \quad \mathrm{O}^{\mathrm{T}} = \mathrm{O}^{-1}:$$

the transpose of an orthogonal matrix is equal to its inverse.

If the elements of the orthogonal matrix are *real*, it is unitary. The reduction described in the preceding paragraph introduced n vectors $\mathbf{e}_1, \mathbf{e}_2, \ldots, \mathbf{e}_n$, which are complex if the corresponding eigenvalues are complex, real if they are real (and therefore equal to ± 1); furthermore, the sum of the squares of the moduli of their components was equal to 1. Let us take one of them, $\mathbf{e}_1$, for example, and suppose it to be complex: $\mathbf{e}_1 = \boldsymbol{\eta}_1 + i\boldsymbol{\eta}_2$, $\boldsymbol{\eta}_1$ and $\boldsymbol{\eta}_2$ being real. Since $\mathbf{e}_1$ is also an eigenvector, we see that $\boldsymbol{\eta}_1$ and $\boldsymbol{\eta}_2$ are two real orthogonal vectors; if we take them to be unit vectors then we have

$$\mathrm{O}\boldsymbol{\eta}_1 = \cos\alpha\,\boldsymbol{\eta}_1 - \sin\alpha\,\boldsymbol{\eta}_2,$$

$$\mathrm{O}\boldsymbol{\eta}_2 = \sin\alpha\,\boldsymbol{\eta}_1 + \cos\alpha\,\boldsymbol{\eta}_2.$$

The matrix O leaves invariant the linear manifold formed by the vectors orthogonal to $\boldsymbol{\eta}_1$ and $\boldsymbol{\eta}_2$; moreover it transforms the vectors of this manifold in an orthogonal manner. It will therefore be possible to find two unit vectors $\boldsymbol{\eta}_3$ and $\boldsymbol{\eta}_4$ orthogonal to each other as well as to $\boldsymbol{\eta}_1$ and $\boldsymbol{\eta}_2$ which transform in an analogous manner with an angle β and so on. To the real eigenvalues ± 1 of the matrix O there will correspond vectors which are reproduced, apart perhaps from a change of sign.

Finally we see that by transforming the matrix O with a real orthogonal matrix C, it can be reduced to a matrix of the form:

$$\begin{pmatrix} \cos\alpha & -\sin\alpha & 0 & 0 & 0 & 0 \\ \sin\alpha & \cos\alpha & 0 & 0 & 0 & 0 \\ 0 & 0 & \cos\beta & -\sin\beta & 0 & 0 \\ 0 & 0 & \sin\beta & \cos\beta & 0 & 0 \\ 0 & 0 & 0 & 0 & 1 & 0 \\ 0 & 0 & 0 & 0 & 0 & -1 \end{pmatrix}.$$

The determinant of such a matrix is equal to ± 1; the matrix is said to be *proper orthogonal* if the determinant is equal to $+1$.

45. Application to the decomposition of a real rotation

The preceding results reveal interesting properties of the group of rotations and reversals in real Euclidean space. In such a space referred to an orthogonal frame of reference the operations are represented by real orthogonal matrices. Let us call a rotation *simple* if, by a suitable choice of orthogonal co-ordinates, the rotation leaves invariant all the co-ordinates except two, x_i and x_j, which transform according to the formulae

$$x'_i = x_i \cos\alpha - x_j \sin\alpha,$$

$$x'_j = x_i \sin\alpha + x_j \cos\alpha;$$

α is the angle of the simple rotation and the biplane formed by $\mathbf{e}_i$ and $\mathbf{e}_j$ is the biplane of the simple rotation. We can then state the following theorem:

THEOREM. *Every rotation is the product of a certain number $\leqslant n/2$ of simple rotations whose biplanes are orthogonal to one another; every reversal is the product of a certain number $\leqslant (n-1)/2$ of simple rotations of the same nature and of a reflection in a hyperplane containing the biplanes of all these simple rotations.*

46. Hermitian matrices

A square matrix H is said to be a *Hermitian* matrix if its transpose is equal to its conjugate complex:

$$H^T = \bar{H}.$$

The transform of a Hermitian matrix H by a unitary matrix U is still Hermitian: this results from the equality

$$(U^{-1}HU)^T = U^T H^T (U^{-1})^T = \bar{U}^{-1}\bar{H}\bar{U}.$$

THEOREM. *A Hermitian matrix H can always be transformed by a unitary matrix in such a way as to make it diagonal with real elements.*

Let us first show that any eigenvalue of H is real. The equation

$$\mathrm{H}x = \lambda x$$

implies, on transposing and taking conjugate complexes of both sides,

$$\bar{x}^{\mathrm{T}}\mathrm{H} = \bar{\lambda}\bar{x}^{\mathrm{T}};$$

multiplying both sides of the first equation on the left by $\bar{x}^{\mathrm{T}}$ and both sides of the second on the right by x, we obtain

$$\bar{x}^{\mathrm{T}}\mathrm{H}x = \lambda\bar{x}^{\mathrm{T}}x = \bar{\lambda}\bar{x}^{\mathrm{T}}x,$$

whence $\bar{\lambda} = \lambda$.

Let us introduce into the vector space in which H operates the unitary metric according to which the scalar product of two vectors $\mathbf{x}$, $\mathbf{y}$ is $\bar{x}^{\mathrm{T}}y$. There exists at least one unit vector $\mathbf{e}_1$, which H reproduces multiplied by a real factor λ_1. Any vector $\mathbf{x}$ orthogonal to $\mathbf{e}_1$ is transformed therefore into a vector $\mathbf{x}^1$ orthogonal to $\mathbf{e}_1$: this results from the fact that

$$\bar{e}_j^{\mathrm{T}}x^1 = \bar{e}_1^{\mathrm{T}}\mathrm{H}x = \bar{\lambda}_1 e_1^{\mathrm{T}}x.$$

The subspace of vectors orthogonal to $\mathbf{e}_1$, being invariant under H, contains at least one unit vector $\mathbf{e}_2$ which H reproduces multiplied by the real factor λ_2. Continuing this reasoning, we arrive at a system of n unit vectors $\mathbf{e}_1, \mathbf{e}_2, \ldots, \mathbf{e}_n$, orthogonal in pairs, which H reproduces multiplied respectively by $\lambda_1, \lambda_2, \ldots, \lambda_n$. We can take these as basis vectors; this amounts to transforming H, by a unitary matrix C, into a diagonal matrix with elements $\lambda_1, \lambda_2, \ldots, \lambda_n$.

47. Comment

The product of two unitary matrices of the same order is a unitary matrix: the product of two orthogonal matrices is an orthogonal matrix. But the product of two Hermitian matrices is not in general a Hermitian matrix. The necessary and sufficient condition for the matrix HH′ to be Hermitian is

$$\mathrm{H}'^{\mathrm{T}}\mathrm{H}^{\mathrm{T}} = \bar{\mathrm{H}}\bar{\mathrm{H}}' \quad \text{or} \quad \mathrm{H}'\mathrm{H} = \mathrm{H}\mathrm{H}';$$

i.e., the two matrices must commute with each other. The set of unitary linear transformations as well as the set of orthogonal transformations thus forms a group, but not the set of Hermitian transformations.

On the other hand, the sum of two Hermitian matrices is a Hermitian matrix; but this property is not possessed by the unitary matrices, nor by the orthogonal matrices.

V. IRREDUCIBILITY OF p-VECTORS

We shall conclude this chapter with a study of multivectors from the point of view of their irreducibility with respect to the group of rotations. We shall confine ourselves to the case of a real Euclidean space referred to an orthogonal frame of reference, but the reasoning can be applied unchanged to the case of a complex space.

48. Irreducibility of a p-vector with respect to the group of rotations ($n \neq 2p$)

We denote by σ_i the reflection associated with the basis vector $\mathbf{e}_i$. If the p-vector is reducible, there will exist a tensor $\mathcal{T}$ of degree $r < {}^nC_p$ which has as its components linear combinations of the components* $x_{i_1 i_2 i_3 \ldots i_p}$ of the p-vector. Assume that in one of these components the coefficient of $x_{123\ldots p}$ is not zero. The rotation $\sigma_1\sigma_2$ will give a new component of $\mathcal{T}$ which can be obtained from the former by changing the signs of the coefficients of these $x_{i_1 i_2 \ldots i_p}$ which have only one of the indices 1 and 2: by addition these quantities can be eliminated. By using each of the rotations $\sigma_1\sigma_3, \ldots, \sigma_1\sigma_p$ in the same way it is possible to obtain a component of $\mathcal{T}$ which contains $x_{12\ldots p}$ and only such quantities $x_{i_1 i_2 \ldots i_p}$ as contain none of the indices $1, 2, \ldots, p$. Finally by using $\sigma_1\sigma_{p+1}, \ldots, \sigma_1\sigma_n$ we can show that there exists a component of $\mathcal{T}$ which besides $x_{12\ldots p}$ only contains those $x_{i_1 i_2 \ldots i_p}$ where all the indices $p + 1$, $p + 2, \ldots, n$ occur, but none of the indices $1, 2, \ldots, p$. This is manifestly impossible unless $n = 2p$.

If then, $n \neq 2p$ and the tensor $\mathcal{T}$ contains the component $x_{12\ldots p}$; then any permutation of the x_i, followed by the necessary change of sign of one of the co-ordinates if the permutation is odd, shows that the tensor $\mathcal{T}$ contains all the components of the p-vector, which is thus irreducible.

49. Semi-ν-vectors in the space $E_{2\nu}$

Assume now that n is even, $n = 2\nu$, and $p = \nu$. The above reasoning shows that the tensor $\mathcal{T}$ contains at least one component of the form

$$x_{12\ldots\nu} + \alpha x_{(+1)(\nu+2\ldots(2\nu)}.$$

A permutation of the indices allows us to deduce from this the existence of the component

$$x_{k_1 k_2 \ldots k_\nu} \pm \alpha x_{k_{\nu+1} k_{\nu+2} \ldots k_{2\nu}},$$

where the + or − sign must be taken according to whether the permutation $(k_1 k_2 \ldots k_{2\nu})$ is even or odd. In particular we will have the component

$$x_{(\nu+1)(\nu+2)\ldots(2\nu)} + (-1)^\nu \alpha x_{12\ldots\nu}.$$

If the tensor $\mathcal{T}$ is to be distinct from the ν-vector, it is necessary that $\alpha^2 = (-1)^\nu$, that is $\alpha = \pm i^\nu$.

Conversely take for example $\alpha = i^\nu$. The quantities

$$x_{k_1 k_2 \ldots k_\nu} + i^\nu x_{k_{\nu+1} k_{\nu+2} \ldots k_{2\nu}},$$

where the permutation $(k_1 k_2 \ldots k_{2\nu})$ is even, generate a tensor. In fact the

*From there on we shall use $x_{i_1 i_2 \ldots i_p}$ to denote the components of a p-vector (in place of the notation $P_{i_1 i_2 \ldots i_p}$ which was introduced above in Section 15).

components of the ν-vector supplementary to the given ν-vector are

$$y_{k_1k_2\ldots k_\nu} = x_{k_{\nu+1}k_{\nu+2}\ldots k_{2\nu}};$$

this supplementary ν-vector forms a tensor equivalent to a ν-vector and it follows that the quantities $x_{k_1k_2\ldots k_\nu} + my_{k_1k_2\ldots k_\nu}$ form a tensor; it is only necessary to give m the value i^ν to obtain the quantities under consideration.

Therefore the ν-vector decomposes into two irreducible tensors; *they are not equivalent*; if they were, the ν-vector would contain an infinity of irreducible tensors of degree ${}^{2\nu}C_\nu/2$ (Section 34), whereas it only contains two.

In a general co-ordinate system, the two tensors, whose existence we have just shown, and which we call semi-ν-vectors of the first and second types, have as components

$$x_{k_1k_2\ldots k_\nu} + \varepsilon i^\nu \sqrt{g}\, x^{k_{\nu+1}k_{\nu+2}\ldots k_{2\nu}} \qquad (\varepsilon = \pm 1).$$

This result follows from the expression for the components of the ν-vector supplementary to a given ν-vector (Section 17).

50. Comments

The preceding results hold for the group of rotations, or even more simply the group of proper rotation in a real pseudo-Euclidean space, even though in this case the proof given needs to be completed. This is a consequence of a general theorem which will be indicated below (Section 75).

We may inquire whether a p-vector and a q-vector ($p \neq q$) can form equivalent tensors. This can only be the case if $q = n - p$, as this is the only case in which the tensors in question are of the same degree.

Under the group of rotations the p-vector and the $(n - p)$-vector are effectively equivalent, since the components of a p-vector and of the corresponding supplementary $(n - p)$-vector are obviously transformed in the same manner by a rotation. On the other hand, we have already seen that if $n = 2\nu$, the semi-ν-vector of the first type and the semi-ν-vector of the second type are not equivalent.

51. The behaviour of multivectors under the group of rotations and reversals

Here the semi-ν-vectors do not form tensors, since a reflection changes a semi-ν-vector of the first type into a semi-ν-vector of the second type.

THEOREM. *The p-vectors ($p = 1, 2, \ldots, n$) form irreducible tensors; no pair of them is equivalent.*

The last part of the theorem is easily proved; it is only necessary to prove that a p-vector and an $(n - p)$-vector ($n \neq 2p$) are not equivalent to each other. If they were, it would be possible to make a correspondence between the components of these two tensors in such a manner that under any rotation or reversal those of the p-vector and those of the $(n - p)$-vector are transformed in the same manner. We know, as far as the group of rotations is concerned, that this can only be done in one way (Section 32): the component y_α of the $(n - p)$-vector which must correspond to a given component $x_{k_1k_2\ldots k_p}$

of the p-vector is, apart from a constant factor, the component of the supplementary $(n - p)$-vector with the same specification, viz. $y^{k_{p+1}k_{p+2}\ldots k_n}$ (the permutation $k_1k_2\ldots k_n$ is assumed to be even) (Section 17). If we take an orthogonal co-ordinate system, then under a reflection corresponding to a basis vector the two components are reproduced one with its sign unchanged, and the other with its sign altered.

CHAPTER III

SPINORS IN THREE-DIMENSIONAL SPACE

I. THE CONCEPT OF A SPINOR

52. Definition

Suppose the three-dimensional space E_3 is referred to a system of orthogonal co-ordinates; let (x_1, x_2, x_3) be an isotropic vector, i.e., have zero length. We can associate with this vector, the components of which satisfy

$$x_1^2 + x_2^2 + x_3^2 = 0,$$

two numbers ξ_0, ξ_1 given by

$$x_1 = \xi_0^2 - \xi_1^2,$$

$$x_2 = i(\xi_0^2 + \xi_1^2),$$

and

$$x_3 = -2\xi_0\xi_1.$$

These equations have two solutions given, for example, by the formulae

$$\xi_0 = \pm\sqrt{\frac{x_1 - ix_2}{2}} \quad \text{and} \quad \xi_1 = \pm\sqrt{\frac{-x_1 - ix_2}{2}}.$$

It is not possible to give a consistent choice of sign which will hold for all isotropic vectors in such a manner that the solution varies continuously with the vector. Thus, suppose there is such a choice; start with a definite isotropic vector and suppose it to be continuously rotated round $0x_3$ through an angle α: $x_1 - ix_2$ will be multiplied by $e^{-i\alpha}$, thus by continuity ξ_0 will be multiplied by $e^{-i\alpha/2}$. When the angle of rotation is 2π, the isotropic vector returns to its original position, but ξ_0 is multiplied by $e^{-i\pi} = -1$; i.e., its value is of the opposite sign to that originally selected.

The pair of quantities ξ_0, ξ_1 constitutes a *spinor*. A spinor is thus a sort of "directed" or "polarised" isotropic vector; a rotation about an axis through an angle 2π changes the polarisation of this isotropic vector.

53. A spinor is a Euclidean tensor

Consider a rotation (or reversal) defined by the equations

$$x'_1 = \alpha x_1 + \beta x_2 + \gamma x_3,$$

$$x'_2 = \alpha' x_1 + \beta' x_2 + \gamma' x_3,$$

$$x'_3 = \alpha'' x_1 + \beta'' x_2 + \gamma'' x_3,$$

where $\alpha, \beta, \gamma, \alpha', \beta', \gamma', \alpha'', \beta'', \gamma''$ are the nine direction cosines of three orthogonal directions. Consider the spinor (ξ_0, ξ_1) associated with an isotropic vector (x_1, x_2, x_3) and one of the spinors (ξ'_0, ξ'_1) associated with the transformed vector; then

$$\begin{aligned}\xi_0'^2 &= \tfrac{1}{2}[(\alpha - i\alpha')x_1 + (\beta - i\beta')x_2 + (\gamma - i\gamma')x_3] \\ &= \tfrac{1}{2}(\alpha - i\alpha' + i\beta + \beta')\xi_0^2 - (\gamma - i\gamma')\xi_0\xi_1 + \tfrac{1}{2}(-\alpha + i\alpha' + i\beta + \beta')\xi_1^2.\end{aligned}$$

Since the discriminant of the quadratic form on the right-hand side is

$$\begin{aligned}(\gamma - i\gamma')^2 - (\alpha - i\alpha' + i\beta + \beta')(-\alpha + i\alpha' + i\beta + \beta') \\ = (\alpha - i\alpha')^2 + (\beta - i\beta')^2 + (\gamma - i\gamma')^2 = 0,\end{aligned}$$

the right-hand side must be a perfect square. Thus the quantity ξ'_0 is linear in ξ_0, ξ_1, and the same is obviously true for ξ'_1. When the sign of ξ'_0 is chosen, the quantity ξ'_1 is determined by the expression

$$\xi'_0\xi'_1 = -\tfrac{1}{2}(\alpha'' + i\beta'')\xi_0^2 + \gamma''\xi_0\xi_1 + \tfrac{1}{2}(\alpha'' - i\beta'')\xi_1^2.$$

The linear substitutions induced in a spinor by rotations and reversals will be discussed later.

54. Geometric meaning of the ratio ξ_0/ξ_1

Let us turn our attention to the ratio ξ_0/ξ_1. This ratio undergoes a homographic transformation under any rotation or reversal. This result is well-known, since ξ_0/ξ_1 can be considered as the parameter of a generator of the isotropic cone, and all rotations and reversals preserve the cross-ratio of any four generators of this cone. Conversely if the homographic transformation is known the rotation can be found. Suppose M is any point in space, then there are two isotropic directions orthogonal to OM; these two directions will transform into two isotropic directions which will be orthogonal to OM^1, this latter direction can thus be found and hence the point M^1. These ideas are the basis of the Euler–Olinde–Rodrigues parameters which will be discussed below.

II. MATRICES ASSOCIATED WITH VECTORS

55. The matrix associated with a vector

Let us return to spinors. The equations for the isotropic direction of the vector associated with a spinor ξ can be written as

$$\left.\begin{aligned} \xi_0 x_3 + \xi_1(x_1 - ix_2) = 0, \\ \xi_0(x_1 + ix_2) - \xi_1 x_3 = 0. \end{aligned}\right\} \qquad (1)$$

This suggests that the matrix

$$\mathrm{X} = \begin{pmatrix} x_3 & x_1 - ix_2 \\ x_1 + ix_2 & -x_3 \end{pmatrix}$$

formed from the coefficients of the left-hand sides of these equations may be of importance. If we regard x_1, x_2, x_3 as the components of a vector $\mathbf{x}$ we say that the matrix X is associated with this vector and we shall often refer to the "vector X" instead of the "vector which is associated with the matrix X".

The matrices X have several remarkable properties:

THEOREM 1. *The determinant of the matrix associated with a vector equals minus the scalar square of the vector.*

THEOREM 2. *The square of the matrix* X *associated with a vector equals the unit matrix multiplied by the scalar square of the vector.*

Proof:

$$\mathrm{XX} = \begin{pmatrix} x_3^2 + (x_1 - ix_2)(x_1 + ix_2) & x_3(x_1 - ix_2) - (x_1 - ix_2)x_3 \\ (x_1 + ix_2)x_3 - x_3(x_1 + ix_2) & (x_1 + ix_2)(x_1 - ix_2) + x_3^2 \end{pmatrix}$$

$$= \begin{pmatrix} x_1^2 + x_2^2 + x_3^2 & 0 \\ 0 & x_1^2 + x_2^2 + x_3^2 \end{pmatrix}.$$

THEOREM 3. *The scalar product of two vectors* $\mathbf{x}, \mathbf{y}$ *equals half the sum of the products* XY *and* YX *of the associated matrices.*

Proof: If λ and μ are two parameters,

$$(\lambda\mathrm{X} + \mu\mathrm{Y})^2 = \lambda^2\mathrm{X}^2 + \mu^2\mathrm{Y}^2 + \lambda\mu(\mathrm{XY} + \mathrm{YX})$$

$$(\lambda\mathbf{x} + \mu\mathbf{y})^2 = \lambda^2\mathbf{x}^2 + \mu^2\mathbf{y}^2 + 2\lambda\mu\mathbf{x}\cdot\mathbf{y}.$$

The theorem follows immediately from these two results.

In particular if we consider the matrices associated with the basis vectors

$$H_1 = \begin{pmatrix} 0 & 1 \\ 1 & 0 \end{pmatrix}, \qquad H_2 = \begin{pmatrix} 0 & -i \\ i & 0 \end{pmatrix}, \qquad H_3 = \begin{pmatrix} 1 & 0 \\ 0 & -1 \end{pmatrix},$$

the squares of these matrices equal 1, and the product of any two of them will change sign if the order of the factors is reversed:

$$H_1^2 = H_2^2 = H_3^2, \quad H_1H_2 = -H_2H_1, \quad H_2H_3 = -H_3H_2, \quad H_3H_1 = -HH_3.$$

56. The matrix associated with a bivector and the matrix associated with a trivector

The bivector given by two vectors **x**, **y** is represented analytically by its components

$$x_2y_3 - x_3y_2, \qquad x_3y_1 - x_1y_3, \qquad x_1y_2 - x_2y_1.$$

The matrix

$$\tfrac{1}{2}(XY - YX) = \begin{pmatrix} i(x_1y_2 - x_2y_1) & i(x_2y_3 - x_3y_2) + (x_3y_1 - x_1y_3) \\ i(x_2y_3 - x_3y_2) - (x_3y_1 - x_1y_3) & -i(x_1y_2 - x_2y_1) \end{pmatrix}$$

can be associated with it.

Note that *this matrix is i times the matrix associated with the vector product of the two given vectors.*

If the two vectors are orthogonal, the matrix associated with the bivector is

$$XY = -YX.$$

In a similar manner the trivector defined by three vectors **x**, **y**, **z** is associated with the matrix

$$\tfrac{1}{6}(XYZ + YZX + ZXY - YXZ - ZYX - XZY).$$

If the three vectors are orthogonal, this matrix reduces to XYZ. If we write **u** for the vector product **x** $\wedge$ **y**, then in this case

$$XYZ = iUZ.$$

Now the product UZ of two matrices associated with two vectors lying in the same direction equals the scalar product of these vectors, and here this equals the volume v of the trivector. *The matrix associated with a trivector of algebraic volume v is thus iv.* In particular

$$H_1H_2H_3 = i,$$

as can be verified by a simple calculation.

57. Relation to the theory of quaternions

There is no linear relation with complex coefficients, not all zero, of the form

$$a_0 + a_1H_1 + a_2H_2 + a_3H_3 = 0.$$

This is easily verified; the left-hand side is the matrix

$$\begin{pmatrix} a_0 + a_3 & a_1 - ia_2 \\ a_1 + ia_2 & a_0 - a_3 \end{pmatrix}$$

It follows that any second order matrix with complex elements can be expressed in one and only one way as the sum of a scalar and a vector. This gives an algebra which is the same as the quaternion algebra. If we write

$$I_1 = -iH_1, \quad I_2 = -iH_2, \quad I_3 = -iH_3,$$

then the laws of multiplication become

$$I_1^2 = I_2^2 = I_3^2 = -1; I_2I_3 = -I_3I_2 = I_1, I_3I_1 = -I_1I_3 = I_2, I_1I_2 = -I_2I_1 = I_3.$$

In the real domain, the matrices 1, i, H_1, H_2, H_3, iH_1, iH_2, iH_3 (which are linearly independent in this domain), form an algebra of order 8 over the field of real numbers. Any element of the algebra is uniquely the sum of a real scalar, a real vector, a real bivector and a real trivector.

III. REPRESENTATIONS OF REFLECTIONS AND ROTATIONS

58. Representation of a reflection

Let **a** be a given unit vector. The vector **x**′, the reflection of a vector **x** in the plane π which passes through the origin and has **a** as normal, is given (Section 9) by

$$\mathbf{x}' = \mathbf{x} - 2\mathbf{a}(\mathbf{x} \cdot \mathbf{a}),$$

or, in terms of matrices (using the result $A^2 = 1$),

$$X' = X - A(XA - AX) = -AXA. \tag{2}$$

It follows that the reflection of a bivector defined by a pair of perpendicular vectors X, Y is given by

$$X'Y' = AXYA,$$

or, in terms of the matrix U associated with the bivector,

$$U' = AUA. \tag{3}$$

Finally, under the same reflection the matrix associated with a trivector is given by a similar formula except for a change of sign.

59. Representation of a rotation

Any rotation can be expressed as the product of a pair of reflections A, B (Section 10); the effect of these, in the case of a vector X and a bivector U, is

$$X' = BAXAB, \qquad U' = BAUAB.$$

Putting S = BA these become

$$X' = SXS^{-1}, \qquad U' = SUS^{-1}. \tag{4}$$

The formulae of Euler–Olinde–Rodrigues can be deduced from (3). Let L be the unit vector along the axis of rotation, and θ the angle of rotation. The two unit-vectors A and B have scalar product $\cos(\theta/2)$ and their vector product $\frac{1}{2}(AB - BA)$ equals $iL\sin(\theta/2)$. From this it follows that

$$BA = \cos\frac{\theta}{2} - iL\sin\frac{\theta}{2}, \qquad AB = \cos\frac{\theta}{2} + iL\sin\frac{\theta}{2},$$

hence

$$X' = \left(\cos\frac{\theta}{2} - iL\sin\frac{\theta}{2}\right)X\left(\cos\frac{\theta}{2} + iL\sin\frac{\theta}{2}\right). \tag{5}$$

Denoting the direction cosines of L by L_1, L_2, L_3 the Euler–Olinde–Rodrigues parameters are the four numbers

$$\rho = \cos\frac{\theta}{2}, \qquad \lambda = l_1\sin\frac{\theta}{2}, \qquad \mu = l_2\sin\frac{\theta}{2}, \qquad \nu = l_3\sin\frac{\theta}{2}.$$

The sum of their squares is unity.

60. Operation on spinors

Returning to spinors, let ξ denote the matrix with two rows and one column with elements ξ_0 and ξ_1. It follows from equation (1) that if the isotropic vector **x** associated with the spinor ξ is also associated with the matrix X, then $X\xi = 0$.

We make the convention that the effect of a reflection in a plane π, normal to the unit vector **a**, is given by the operation

$$\xi' = A\xi. \tag{6}$$

We must show that this convention is consistent with our previous results.

If X′ is the matrix associated with **x**′, the reflection of **x** in the plane π, then

$$X'\xi' = -(AXA)A\xi = -AX\xi = 0,$$

i.e., the spinor associated with **x**′ is of the form $m\xi'$.

In the special case where **a** is the basis vector $\mathbf{e}_3$:

$$A = H_3 = \begin{pmatrix} 1 & 0 \\ 0 & -1 \end{pmatrix},$$

formula (6) gives

$$\xi'_0 = \xi_0, \qquad \xi'_1 = -\xi_1;$$

the spinor ξ' is in fact one of the spinors associated with X′, the reflection of X. To see that this is true in the general case, we note that when the operation $\xi \to mA\xi$ is repeated twice we must return to a spinor associated with the original isotropic vector, i.e., ξ or $-\xi$; as a consequence $m^2 = \pm 1$, but m varies continuously with A and must thus be constant. Thus the convention is consistent.

Note that any reflection when applied to a spinor can lead to either of two spinors given by $\xi' = A\xi$ and $\xi' = -A\xi$; the actual transformation depends on the choice of unit vector normal to the plane of reflection.

A rotation, which is the product of two reflections, is similarly two-valued. The effect of the rotation resulting from the reflection A followed by reflection B is given by

$$\xi' = BA\xi. \tag{7}$$

Geometrically this rotation is the same as $-BA$ which gives the operation

$$\xi' = -BA\xi.$$

IV. THE PRODUCT OF TWO SPINORS AND ITS DECOMPOSITION INTO IRREDUCIBLE PARTS

The three monomials ξ_0^2, $\xi_0\xi_1$, and ξ_1^2 *a priori* form a tensor which is equivalent to a vector; they are in fact linearly independent combinations of the components of an isotropic vector, but an isotropic vector, considered as a tensor, is equivalent to a general vector.

61. The matrix

Consider now the fourth order tensor $\xi_\alpha\xi'_\beta$ formed as a "product" of two spinors $\boldsymbol{\xi}$ and $\boldsymbol{\xi}'$. We introduce here the matrix

$$C = \begin{pmatrix} 0 & 1 \\ -1 & 0 \end{pmatrix}, \tag{8}$$

which has the following remarkable property, that if X is any vector whatever,

$$CX = -X^T C. \tag{9}$$

Also

$$C^T = -C, \qquad CC^T = 1, \qquad C^2 = -1. \tag{10}$$

The result (9) is easily verified from the general expression for X (Section 55): thus

$$CX = \begin{pmatrix} 0 & 1 \\ -1 & 0 \end{pmatrix}\begin{pmatrix} x_3 & x_1 - ix_2 \\ x_1 + ix_2 & -x_3 \end{pmatrix} = \begin{pmatrix} x_1 + ix_2 & -x_3 \\ -x_3 & -x_1 + ix_2 \end{pmatrix},$$

$$X^T C = \begin{pmatrix} x_3 & x_1 + ix_2 \\ x_1 - ix_2 & -x_3 \end{pmatrix}\begin{pmatrix} 0 & 1 \\ -1 & 0 \end{pmatrix} = \begin{pmatrix} -x_1 - ix_2 & x_3 \\ x_3 & x_1 - ix_2 \end{pmatrix}.$$

62. The trivector and the vector associated with a pair of spinors

Consider two spinors ξ and ξ', and the quantity $\xi'^{T}C\xi$. Under the reflection A, this quantity becomes, using equations (6) and (9),

$$\xi'^{T}A^{T}CA\xi = -\xi'^{T}CA^{2}\xi = -\xi'^{T}C\xi,$$

i.e., it is unchanged except for its sign. It is thus a tensor which is equivalent to a trivector; it is invariant under rotations, and changes sign under a reversal. Its explicit value is

$$\xi'_{0}\xi_{1} - \xi'_{1}\xi_{0}.$$

It is clear *a priori* that such a quantity when transformed by a linear substitution has its value multiplied by the determinant of the substitution; thus the determinant of A is indeed -1.

Let us now consider, at the same time as the two spinors, an arbitrary vector X, and study the effect of the reflection A on the quantity

$$\xi'^{T}CX\xi.$$

Using Equations (2), (6), and (9), it is seen to transform into

$$-(\xi'^{T}A)C(AXA)A\xi = \xi'^{T}CX\xi,$$

i.e., it is invariant under all rotations and reversals. Now this expression is a bilinear form in the components x_1, x_2, x_3 of the vector X, and the products $\xi\xi'_{\beta}$. It follows (Section 27) that the coefficients of x_1, x_2 and x_3 form a tensor; this tensor must be equivalent to a vector; since the sum $x_1y_1 + x_2y_2 + x_3y_3$ where y_1, y_2, y_3 are the components of a vector, is also invariant under all rotations and reversals, the tensor in question is equivalent to a tensor y_2. The components of the vector thus defined can be expressed as symmetric functions of the components of ξ and of ξ'; since $\xi^{T}CX\xi'$ is a scalar we have

$$\xi^{T}CX\xi' = \xi'^{T}X^{T}C^{T}\xi = -\xi'^{T}X^{T}C\xi = \xi'^{T}CX\xi.$$

The actual values of these components are

$$y_1 = \xi'^{T}CH_1\xi = \xi'_{0}\xi_{0} - \xi'_{1}\xi_{1}$$

$$y_2 = \xi'^{T}CH_2\xi = i\xi'_{0}\xi_{0} + i\xi_{1}\xi_{1}$$

$$y_3 = \xi'^{T}CH_3\xi = -\xi'_{0}\xi_{1} - \xi'_{1}\xi_{0}.$$

When $\xi' = \xi$ this reduces to the isotropic vector from which the spinor ξ was formed; note that when $\xi' \neq \xi$ the scalar square of this vector is equal to $(\xi'_{0}\xi_{1} - \xi'_{1}\xi_{0})^{2}$.

Thus the tensor $\xi_{\alpha}\xi'_{\beta}(\alpha, \beta = 0, 1)$ *of degree* 4 *has been decomposed into a vector and a trivector, the volume of the trivector being equal to the length of the vector.*

V. CASE OF REAL EUCLIDEAN SPACE

All the above results apply in the domain of complex rotations and reversals. We now confine the discussion to real Euclidean space and consider only real rotations and reversals.

63. Complex conjugate vectors

The matrices X and Y associated with two complex conjugate vectors are

$$X = \begin{pmatrix} x_3 & x_1 - ix_2 \\ x_1 + ix_2 & -x_3 \end{pmatrix}, \qquad Y = \begin{pmatrix} \bar{x}_3 & \bar{x}_1 - i\bar{x}_2 \\ \bar{x}_1 + i\bar{x}_2 & -\bar{x}_3 \end{pmatrix};$$

obviously

$$Y = \bar{X}^T \quad \text{or} \quad \bar{Y} = X^T. \tag{11}$$

In particular if $\mathbf{x}$ is real, $\bar{X} = X^T$ which proves the following theorem.

THEOREM. *A real vector is associated with a Hermitian matrix.*

The matrix U associated with a bivector, being the product of i with the matrix associated with a real vector, is not Hermitian; in fact

$$\bar{U} = -U^T.$$

Any rotation is represented by a matrix S = BA, the product of the matrices associated with two real unit vectors; it follows that

$$\bar{S}^T = \bar{A}^T\bar{B}^T = AB = (BA)^{-1} = S^{-1},$$

which proves the following theorem.

THEOREM. *Any rotation is represented by a unitary matrix of unit determinant; any reversal is represented by a unitary matrix of determinant* -1.

64. Conjugate spinors

If X is an isotropic vector and ξ is one of the spinors associated with X, then $X\xi = 0$. Thus

$$C\bar{X}\bar{\xi} = 0,$$

and using (9) and (11)

$$0 = C\bar{X}\bar{\xi} = -\bar{X}^T C\bar{\xi} = -YC\bar{\xi},$$

where Y is the complex conjugate vector to X. It follows that each of the spinors associated with Y is of the form $mC\bar{\xi}$, where the coefficient m can easily be shown to equal $\pm i$.

For definiteness we define the spinor conjugate to ξ as $iC\bar{\xi}$. The operation of conjugation is not an involution, since on repeating it the result is $-\xi$.

It is important to note that the spinor $iC\bar{\xi}$ is actually a different sort of quantity from ξ; under the reflection A the matrix $iC\bar{\xi}$ becomes $iC\bar{A}\bar{\xi} = iCA^T\bar{\xi} = -iA(C\bar{\xi})$: i.e., it becomes multiplied on the left by $-A$ not by A. The conjugate of a spinor will be called a "spinor of the second type".

65. The scalar and bivector associated with two conjugate spinors

If in the product of two spinors ξ and ξ', ξ' is replaced by the conjugate of ξ, i.e., by $C\bar{\xi}$, the tensor thus obtained can be decomposed into two irreducible parts. One is given by the quantity

$$(\bar{\xi}^T C^T) C\xi = \bar{\xi}^T \xi = \xi_0 \bar{\xi}_0 + \xi_1 \bar{\xi}_1;$$

this is a *scalar* which is invariant under all rotations and reversals; we have already shown (Section 63) that any rotation or reversal is represented by a unitary matrix. The other irreducible part is given by the quantity

$$(\bar{\xi}^T C^T) CX\xi = \bar{\xi}^T X\xi = (x_1 + ix_2)\xi_0\bar{\xi}_1 + (x_1 - ix_2)\xi_1\bar{\xi}_0 + x_3(\xi_0\bar{\xi}_0 - \xi_1\bar{\xi}_1).$$

The coefficients of x_1, x_2 and x_3 form a bivector. Under a real reflection A the quantity $\bar{\xi}^T X\xi$ becomes

$$-(\bar{\xi}^T \bar{A}^T)(AXA)A\xi = -\bar{\xi}^T X\xi:$$

it is unchanged under rotations, but changes sign under reversals; it is thus equivalent to a trivector. But if y_{23}, y_{31}, y_{12} are the components of a bivector, then the quantity $x_1 y_{23} + x_2 y_{3}, + x_3 y_{12}$ is a trivector. Thus using Theorem II of Section 27, *the three quantities*

$$\xi_0\bar{\xi}_1 + \xi_1\bar{\xi}_0, \qquad i(\xi_0\bar{\xi}_1 - \xi_1\bar{\xi}_0), \qquad \xi_0\bar{\xi}_0 - \xi_1\bar{\xi}_1$$

are the components of a bivector. That they are not components of a vector follows immediately from the result that under a reflection in the origin they are unchanged, since ξ_0 and ξ_1 both become multiplied by i.

VI. CASE OF PSEUDO–EUCLIDEAN SPACE

66. Real Rotations

A pseudo–Euclidean space can be obtained by replacing x_2 by ix_2 in all expressions, and taking the new co-ordinates x_1, x_2, x_3 as real. The matrix associated with a vector is now real:

$$X = \begin{pmatrix} x_3 & x_1 - x_2 \\ x_1 + x_2 & -x_3 \end{pmatrix}.$$

An isotropic vector (light-vector) with positive time-like component ($x_2 > 0$) is associated with two spinors with real components:

$$x_1 = \xi_0^2 - \xi_1^2, \qquad x_2 = \xi_0^2 + \xi_1^2, \qquad x_3 = -2\xi_0\xi_1.$$

The conjugate of a spinor ξ is $\bar{\xi}$; this is of the same type as ξ.

The expression $\bar{\xi}^{\mathrm{T}}\mathrm{C}\xi$ defines a trivector $\bar{\xi}_0\xi_1 - \bar{\xi}_1\xi_0$. The expression $\bar{\xi}^{\mathrm{T}}\mathrm{CX}\xi$ defines a real vector

$$y_1 = \xi_0\bar{\xi}_0 - \xi_1\bar{\xi}_1, \qquad y_2 = \xi_0\bar{\xi}_0 + \xi_1\bar{\xi}_1, \qquad y_3 = -(\xi_0\bar{\xi}_1 + \xi_1\bar{\xi}_0);$$

it is a time-like vector of length $i(\xi_0\bar{\xi}_1 - \xi_1\bar{\xi}_0)$; it is isotropic if the spinor ξ is real.

Finally, *proper* rotations are represented by real matrices with unit determinant, and *proper* reversals by real matrices with determinant -1.

CHAPTER IV

LINEAR REPRESENTATIONS OF THE GROUP OF ROTATIONS IN E_3

I. LINEAR REPRESENTATIONS GENERATED BY SPINORS

A simple construction will be described, by means of which an unlimited series of irreducible linear representations of either the group of rotations, or the group of rotations and reversals, in real or complex three-dimensional Euclidean space, can be generated. It will also be proved that, at least in real space, all such representations can be so constructed, i.e., any other representation of these groups is completely reducible. It follows from a theorem, stated below, that any linear representation whatsoever of the real groups provides a representation of the corresponding complex group, and that these two representations are either both reducible, or both irreducible.

67. The representation $\mathscr{D}_{p/2}$ and its generating polynomial

We start from the spinor (ξ_0, ξ_1). The set of homogeneous polynomials of degree p in ξ_0 and ξ_1 constitutes a tensor and thus forms a linear representation of the rotation group. This tensor can be represented symbolically by the *generating polynomial* $(a\xi_0 + b\xi_1)^p$, where a and b are two arbitrary parameters, i.e., the coefficients of the different monomials in a and b in the expansion of this polynomial are the components of the tensor. We denote the tensor, or the corresponding linear representation, by $\mathscr{D}_{p/2}$.

THEOREM. *The representation $\mathscr{D}_{p/2}$ is irreducible.*

It is only necessary to prove this for the group of complex rotations. A rotation through a real angle θ about the x_3 axis is given by the operation

$$\xi_0' = \xi_0 e^{i(\theta/2)}, \qquad \xi_1' = \xi_1 e^{-i(\theta/2)}.$$

Under this operation $\xi_0^\alpha \xi_1^\beta$ is multiplied by $e^{i(\alpha-\beta)(\theta/2)}$. Unless θ has a special value, the multipliers corresponding to different components $\xi_0^\alpha \xi_1^\beta$ of the tensor will be distinct. Assume that the representation $\mathscr{D}_{p/2}$ is not irreducible. Then there must be a tensor $\mathscr{T}$ formed from $q < p + 1$ polynomials of degree p in ξ_0 and ξ_1; let $A_0\xi_0^p + A_1\xi_0^{p-1}\xi_1 + \cdots + A_p\xi^p$ be one of these polynomials. The rotation described above transforms this into

$$A_0 e^{i(p/2)\theta}\xi_0^p + A_1 e^{i[(p/2)-1]\theta}\xi_0^{p-1}\xi_1 + \cdots + A_p$$

On repeating this operation p times, $p + 1$ linearly independent combinations of

$$A_0\xi_0^p, \qquad A_1\xi_0^{p-1}\xi_1, \qquad A_2\xi_0^{p-2}\xi_1^2, \ldots, \qquad A_p\xi_1^p$$

are obtained. From this it follows that there is at least one monomial $\xi_0^h\xi_1^k$ $(h + k = p)$ contained in the tensor $\mathscr{T}$, and thus that all polynomials $(\alpha\xi_0 + \beta\xi_1)^h(\gamma\xi_0 + \delta\xi_1)^k$ with $\alpha\delta - \beta\gamma = 1$ also are included in $\mathscr{T}$. If we take the four constants α, β, γ and δ as all non-zero, then the coefficient of ξ_0^p in the corresponding polynomial is not zero. Thus the tensor $\mathscr{T}$ includes ξ_0^p, and therefore also $(a\xi_0 + b\xi_1)^p$, i.e., all the monomials $\xi_0^h\xi_1^k$.

For $p = 0$, a tensor of degree 1 is obtained, i.e., a scalar; for $p = 1$, a spinor; and for $p = 2$, a vector: the quantities ξ_0^2, $\xi_0\xi_1$, ξ_1^2, in fact, after a suitable linear substitution, transform as do the components of a vector.

There is another representation of degree $p + 1$ of the group of rotations *and reversals*. This is formed from the same components, but under reversals they undergo the above substitution followed by a change in the sign of all the components. This representation will be denoted by $\mathscr{D}_{p/2}^-$, and the first representation will be denoted by $\mathscr{D}_{p/2}^+$; by an argument given above they are not equivalent. This holds for $\mathscr{D}_{\frac{1}{2}}^-$ which gives a spinor of the second type, $\mathscr{D}_1^-$ a bivector, and $\mathscr{D}_0^-$ a trivector (i.e., $\xi_0\xi_1' - \xi_1\xi_0'$). The generating polynomial of $\mathscr{D}_{p/2}^-$ can be taken as

$$(\xi_0\xi_1' - \xi_1\xi_0')(a\xi_0 + b\xi_1)^p.$$

68. Reduction of $\mathscr{D}_i \times \mathscr{D}_j$

Let $u_1, u_2, \ldots, u_{2i+1}$ be the variables of the linear representation $\mathscr{D}_i$. Let $v_1, v_2, \ldots, v_{2j+1}$ be the variables of the other linear representation $\mathscr{D}_j$. The products $u_\alpha v_\beta$ will lead to a new linear representation of degree $(2i + 1)(2j + 1)$ which we will denote by $\mathscr{D}_i \times \mathscr{D}_j$. In general this representation is not irreducible; its reduction into irreducible representations will now be given.

First, note that if ξ and ξ' denote two arbitrary spinors, the generating polynomials

$$P = (a\xi_0 + b\xi_1)^p(a\xi_0' + b\xi_1')^q,$$

$$Q = (a\xi_0 + b\xi_1)^p(a\xi_0 + b\xi_1)^q = (a\xi_0 + b\xi_1)^{p+q}$$

define two equivalent representations. This follows from the fact that under any rotation or reversal the coefficients of the different monomials $a^\alpha b^\beta$ in the two polynomials transform into each other in the same manner, since ξ_0' and ξ_1' transform in the same way as ξ_0 and ξ_1.

We take for the components of $\mathscr{D}_{p/2} \times \mathscr{D}_{q/2}$ polynomials in $\xi_0, \xi_1, \xi'_0, \xi'_1$ which are of degree p in ξ_0, ξ_1 and of degree q in ξ'_0, ξ'_1. Consider the polynomials

$$\phi_0 \equiv (a\xi_0 + b\xi_1)^p(a\xi'_0 + b\xi'_1)^q$$
$$\phi_1 \equiv (\xi_0\xi'_1 - \xi_1\xi'_0)(a\xi_0 + b\xi_1)^{p-1}(a\xi'_0 + b\xi'_1)^{q-1}$$
$$\phi_2 \equiv (\xi_0\xi'_1 - \xi_1\xi'_0)^2(a\xi_0 + b\xi_1)^{p-2}(a\xi'_0 + b\xi'_1)^{q-2}$$
$$\cdots\cdots\cdots\cdots\cdots\cdots\cdots\cdots\cdots\cdots$$
$$\phi_q \equiv (\xi_0\xi'_1 - \xi_1\xi'_0)^q(a\xi_0 + b\xi_1)^{p-q}$$

where we take $p \geqslant q$. Each one of these defines an irreducible linear representation, viz.,

$$\mathscr{D}^+_{(p+q)/2}, \qquad \mathscr{D}^-_{(p+q)/2-1}, \qquad \mathscr{D}^+_{(p+q)/2-2}, \ldots, \qquad \mathscr{D}^{\pm}_{(p-q)/2};$$

the last one has index + or index − depending upon whether q is even or odd.

The components of each of these representations are components of the given representation; the total number of components is

$$(p + q + 1) + (p + q - 1) + (p + q - 3) + \cdots + (p - q + 1)$$
$$= (q + 1)(p + q + 1) - q(q + 1) = (q + 1)(p + 1);$$

which is just the degree of the given representation. These components are linearly independent, otherwise the reducible tensor $\mathscr{D}_{p/2} \times \mathscr{D}_{q/2}$ would have all the components of at least one of its q irreducible tensors zero (Section 34), and this is not so. We have thus proved the following theorem.

THEOREM. *The product of two irreducible linear representations $\mathscr{D}^+_{p/2}$, $\mathscr{D}^+_{q/2}$ is completely reducible and can be decomposed into the irreducible representations*

$$\mathscr{D}^+_{(p+q)/2}, \qquad \mathscr{D}^-_{(p+q)/2-1}, \qquad \mathscr{D}^+_{(p+q)/2-2}, \ldots, \qquad \mathscr{D}^{\pm}_{(p-q)/2}.$$

69. Special cases; Harmonic polynomials

The case of those representations $\mathscr{D}_i$ where the index i is an integer ($p = 2i$) can be discussed separately. For $i = 1$, the generating polynomial

$$(a\xi_0 + b\xi_1)^2 = a^2\xi_0^2 + 2ab\xi_0\xi_1 + b^2\xi_1^2$$

can be replaced by another expression linear in the three components x_1, x_2, x_3 of a vector; thus

$$\xi_0^2 \sim -x_1 + ix_2, \qquad \xi_0\xi_1 \sim x_3, \qquad \xi_1^2 \sim x_1 + ix_2,$$

where the symbol $\sim$ means that the three terms on the right-hand sides transform in the same manner as those on the left-hand sides. It follows that the generating polynomial for $\mathscr{D}_1$ is

$$(b^2 - a^2)x_1 + i(b^2 + a^2)x_2 + 2abx_3.$$

The generating polynomial for $\mathscr{D}_p$ can thus be taken to be

$$F_p \equiv [(b^2 - a^2)x_1 + i(b^2 + a^2)x_2 + 2abx_3]^p.$$

The corresponding tensor has as its components $2p + 1$ homogeneous polynomials of degree p in x_1, x_2, x_3; these are the *harmonic polynomials.* A calculation shows that

$$\frac{\partial^2 F}{\partial x_1^2} + \frac{\partial^2 F}{\partial x_2^2} + \frac{\partial^2 F}{\partial x_3^2} = 0.$$

The product $\mathscr{D}_1^+ \times \mathscr{D}_1^+$, i.e., the product of two vectors, reduces to the three irreducible representations $\mathscr{D}_2^+, \mathscr{D}_1^-$ and $\mathscr{D}_0^+$. The first corresponds to the tensor

$$x_1x_1' - x_3x_3', x_2x_2' - x_3x_3', x_2x_3' + x_3x_2', x_3x_1' + x_1x_3', x_1x_2' + x_2x_1',$$

which is equivalent to the tensor

$$x_1^2 - x_3^2, \quad x_2^2 - x_3^2, \quad 2x_2x_3, \quad 2x_3x_1, \quad 2x_1x_2,$$

formed by the harmonic polynomials of degree two. The second corresponds to the bivector

$$x_2x_3' - x_3x_2', \quad x_3x_1' - x_1x_3', \quad x_1x_2' - x_2x_1'$$

and the third corresponds to the scalar

$$x_1x_1' + x_2x_2' + x_3x_3'.$$

70. Applications

Consider a vector field V_1, V_2, V_3; i.e., with each point (x_1, x_2, x_3) of space is associated a vector with components V_1, V_2, V_3 which are given functions of (x_1, x_2, x_3). Under a rotation or reversal the quantities $\partial V_i/\partial x_j$ behave like the product of two vectors each drawn from the origin, viz.,

$$\frac{\partial}{\partial x_i}, \frac{\partial}{\partial x_2}, \frac{\partial}{\partial x_3} \quad \text{and} \quad V_1, V_2, V_3.$$

We seek all linear functions of the $\partial V_i/\partial x_j$ with constant coefficients which are unchanged by rotations or reversals. It is merely necessary to write the $\partial V_i/\partial x_j$ as a product of two (symbolic) vectors, as above, then to reduce this product, and finally to equate to zero each of the components in one or more of the irreducible tensors. If we consider just one of the resulting tensors, then we obtain one of the following sets:

(a) $$\frac{\partial V_1}{\partial x_1} = \frac{\partial V_2}{\partial x_2} = \frac{\partial V_3}{\partial x_3}, \quad \frac{\partial V_2}{\partial x_3} + \frac{\partial V_3}{\partial x_2} = 0, \quad \frac{\partial V_3}{\partial x_1} + \frac{\partial V_1}{\partial x_3} = 0,$$

$$\frac{\partial V_1}{\partial x_2} + \frac{\partial V_2}{\partial x_1} = 0.$$

(b) $$\frac{\partial V_2}{\partial x_3} - \frac{\partial V_3}{\partial x_2} = 0, \quad \frac{\partial V_3}{\partial x_1} - \frac{\partial V_1}{\partial x_3} = 0, \quad \frac{\partial V_1}{\partial x_2} - \frac{\partial V_2}{\partial x_1} = 0.$$

(c) $$\frac{\partial V_1}{\partial x_1} + \frac{\partial V_2}{\partial x_2} + \frac{\partial V_3}{\partial x_3} = 0.$$

Case (c) gives a vector field of zero divergence;
(b) gives an irrotational vector field (i.e., zero curl);
(a) gives the velocity field for a rigid body displacement.

If the group of rotations only is considered, then since $\mathscr{D}_1^+$ and $\mathscr{D}_1^-$ transform in the same way, the system of equations

$$\frac{\partial V_2}{\partial x_3} - \frac{\partial V_3}{\partial x_2} = mV_1, \qquad \frac{\partial V_3}{\partial x_1} - \frac{\partial V_1}{\partial x_3} = mV_2, \qquad \frac{\partial V_1}{\partial x_2} - \frac{\partial V_2}{\partial x_1} = mV_3,$$

where m is a constant, remains invariant; under a rotation followed by a reflection, each equation transforms into a similar one, with m replaced by $-m$. This system of equations implies that the divergence of the vector field is zero.

71. Dirac's equations

In the same way we can consider the tensor $(\partial\xi_0/\partial x_i, \partial\xi_1/\partial x_i)$ which is transformed under a rotation or a reversal as a product of a vector $\partial/\partial x$ and a spinor ξ. That is, as $\mathscr{D}_1^+ \times \mathscr{D}_{\frac{1}{2}}^+ = \mathscr{D}_{\frac{3}{2}}^+ + \mathscr{D}_{\frac{1}{2}}^-$. The generating polynomial of $\mathscr{D}_1^+ \times \mathscr{D}_{\frac{1}{2}}^+$ is $(a\xi_0' + b\xi_1')^2(a'\xi_0 + b'\xi_1)$, that of $\mathscr{D}_{\frac{3}{2}}^+$ is

$$(a\xi_0' + b\xi_1')^2(a\xi_0 + b\xi_1)$$

$$\sim \left[(b^2 - a^2)\frac{\partial}{\partial x_1} + i(b^2 + a^2)\frac{\partial}{\partial x_2} + 2ab\frac{\partial}{\partial x_3}\right](a\xi_0 + b\xi_1)$$

$$\sim -a^3\left(\frac{\partial\xi_0}{\partial x_1} - i\frac{\partial\xi_0}{\partial x_2}\right) + a^2b\left(2\frac{\partial\xi_0}{\partial x_3} - \frac{\partial\xi_1}{\partial x_1} + i\frac{\partial\xi_1}{\partial x_2}\right)$$

$$+ ab^2\left(2\frac{\partial\xi_1}{\partial x_3} + \frac{\partial\xi_0}{\partial x_1} + i\frac{\partial\xi_0}{\partial x_2}\right) + b^3\left(\frac{\partial\xi_1}{\partial x_1} + i\frac{\partial\xi_1}{\partial x_2}\right):$$

and that of $\mathscr{D}_{\frac{1}{2}}^-$ is

$$(\xi_0'\xi_1 - \xi_1'\xi_0)(a\xi_0' + b\xi_1') = a(-\xi_0'\xi_1'\xi_0 + \xi_0'^2\xi_1) + b(-\xi_1'^2\xi_0 + \xi_0'\xi_1'\xi_1)$$

$$\sim a\left(-\frac{\partial\xi_0}{\partial x_3} - \frac{\partial\xi_1}{\partial x_1} + i\frac{\partial\xi_1}{\partial x_2}\right)$$

$$+b\left(-\frac{\partial\xi_0}{\partial x_1} - i\frac{\partial\xi_0}{\partial x_2} + \frac{\partial\xi_1}{\partial x_3}\right).$$

The equations obtained by equating the components of the tensor $\mathscr{D}_{\frac{1}{2}}^-$ to zero, viz.,

$$\frac{\partial\xi_0}{\partial x_3} + \frac{\partial\xi_1}{\partial x_1} - i\frac{\partial\xi_1}{\partial x_2} = 0,$$

$$\frac{\partial\xi_0}{\partial x_1} + i\frac{\partial\xi_0}{\partial x_2} - \frac{\partial\xi_1}{\partial x_3} = 0,$$

belong to the type of *Dirac equations*; they are the most simple examples of this type.

The equations

$$\frac{\partial \xi_0}{\partial x_3} + \frac{\partial \xi_1}{\partial x_1} - i\frac{\partial \xi_1}{\partial x_2} = m\xi_0$$

$$\frac{\partial \xi_0}{\partial x_1} + i\frac{\partial \xi_0}{\partial x_2} - \frac{\partial \xi_1}{\partial x_3} = m\xi_1$$

are unchanged by a rotation, since spinors of the first and second types $\mathscr{D}_{\frac{1}{2}}^{+}$ and $\mathscr{D}_{\frac{1}{2}}^{-}$ transform in the same way under rotations; under a reflection these equations transform into similar equations with the sign of the constant m altered.

Note that the Dirac equations can also be written symbolically as $(\partial/\partial x)\xi = 0$ where $\partial/\partial x$ is the matrix associated with the vector $\partial/\partial x_1$, $\partial/\partial x_2$, $\partial/\partial x_3$. On multiplying on the left by $\partial/\partial x$, and noting that the square of $\partial/\partial x$ is $\nabla^2 = \partial^2/\partial x_1^2 + \partial^2/\partial x_2^2 + \partial^2/\partial x_3^2$, the equations $\nabla^2\xi_0 = 0$, $\nabla^2\xi_1 = 0$ are obtained.

The equations given by equating the components of $\mathscr{D}_{\frac{3}{2}}^{+}$ to zero, viz.,

$$\frac{\partial \xi_0}{\partial x_1} - i\frac{\partial \xi_0}{\partial x_2} = 0, \qquad \frac{\partial \xi_1}{\partial x_1} + i\frac{\partial \xi_1}{\partial x_2} = 0, \qquad 2\frac{\partial \xi_0}{\partial x_3} - \frac{\partial \xi_1}{\partial x_1} + i\frac{\partial \xi_1}{\partial x_2} = 0,$$

$$2\frac{\partial \xi_1}{\partial x_3} + \frac{\partial \xi_0}{\partial x_1} + i\frac{\partial \xi_0}{\partial x_2} = 0$$

give

$$\xi_0 = b(-x_1 + ix_2) + ax_3 + h, \qquad \xi_1 = a(x_1 + ix_2) + bx_3 + k$$

where a, b, h, and k are four arbitrary constants.

II. INFINITESIMAL ROTATIONS AND THE DETERMINATION OF EUCLIDEAN TENSORS

72. Infinitesimal rotations in the space E_3

We now undertake the search for all linear representations of the rotation group. We have already considered (Section 19) infinitesimal rotations, which, when applied to vectors, define a velocity field for the motion of a rigid body with one point fixed. The most general infinitesimal rotation when applied to a vector x_1, x_2, x_3 is represented by a third order matrix which is a linear combination (with real or complex coefficients) of three basis matrices, e.g., those which represent rotations with unit angular velocity about the three co-ordinate axes; these matrices are

$$\begin{pmatrix} 0 & 0 & 0 \\ 0 & 0 & -1 \\ 0 & 1 & 0 \end{pmatrix}, \quad \begin{pmatrix} 0 & 0 & 1 \\ 0 & 0 & 0 \\ -1 & 0 & 0 \end{pmatrix}, \quad \begin{pmatrix} 0 & -1 & 0 \\ 1 & 0 & 0 \\ 0 & 0 & 0 \end{pmatrix}.$$

When applied to spinors, rotations are represented by analogous matrices. A rotation θ about $\mathbf{e}_3$ is represented, as shown above (Section 59) by the matrix

$$\cos\frac{\theta}{2} - i\mathrm{H}_3 \sin\frac{\theta}{2} = 1 - \frac{i}{2}\theta\mathrm{H}_3 + \cdots.$$

A rotation of unit angular velocity is thus represented by $-\frac{1}{2}i\mathrm{H}_3$. The required matrices are thus

$$-\tfrac{1}{2}i\mathrm{H}_1 = -\frac{1}{2}\begin{pmatrix} 0 & i \\ i & 0 \end{pmatrix}, \qquad -\tfrac{1}{2}i\mathrm{H}_2 = -\frac{1}{2}\begin{pmatrix} 0 & -1 \\ 1 & 0 \end{pmatrix}, \qquad -\tfrac{1}{2}i\mathrm{H}_3 = \frac{1}{2}\begin{pmatrix} i & 0 \\ 0 & -i \end{pmatrix}.$$

73. Definition of the representation to be determined

Consider now any linear representation whatsoever of the rotation group, not necessarily one-valued. It is important to define exactly what is meant by this.

Take the set of rotations in a sufficiently small neighbourhood of the identity rotation, e.g., all rotations through angles less than some fixed angle $\alpha \leqslant \pi$. To each such rotation $\mathscr{R}$ in this neighbourhood we assign one, and only one, matrix S of the given order which satisfies the following conditions:

(i) The elements of S are continuous functions of the parameters which define $\mathscr{R}$.

(ii) If $\mathscr{R}$, $\mathscr{R}'$ and $\mathscr{R}\mathscr{R}'$ belong to the given neighbourhood, the product SS′ of the matrices which correspond to $\mathscr{R}$ and $\mathscr{R}'$ is equal to the matrix which corresponds to $\mathscr{R}\mathscr{R}'$.

Since a rotation through any angle can be obtained as the product of a finite number of rotations in the given neighbourhood, it can be made to correspond to the product matrix of the matrices corresponding to these separate rotations. This correspondence gives what is known as a representation of the rotation group. The matrices which operate as spinors and which we have put into correspondence with rotations satisfy the above conditions, the angle α being equal to π.

If the representation is not one-valued, then, even though S varies continuously as its corresponding rotation is varied continuously, when the rotation returns to its original value, the matrix S does not return to its original value.

74. A fundamental theorem

The linear substitutions associated with a linear representation form a linear group in which the elements of the representation matrices are continuous functions of the parameters which define the rotation. We can prove the following fundamental theorem*:

* This theorem was given by J. von Neumann; it is a special case of a more general theorem due to E. Cartan ("La théorie des groupes finis et continus et l'Analysis situs", *Mém. Sc. Math.*, XLII, 1930, **26**, p. 22.)

THEOREM. *Any continuous linear group can be generated from its infinitesimal elements.*

In the particular case we are dealing with, this means that if the rotation $\mathscr{R}$ is made round the $\mathbf{e}_i$ axis with angle of rotation $\theta_i < \alpha$, the corresponding matrix S has the property that as θ_i tends to zero, $S - 1/\theta_i$ tends to a definite matrix R_i which represents a rotation of unit angular velocity round $\mathbf{e}_i$. In general the infinitesimal rotation with unit angular velocity about an axis with direction cosines α, β, γ is represented by the matrix

$$\alpha R_1 + \beta R_2 + \gamma R_3.$$

Finally if we imagine a continuous series of rotations depending on a parameter t and if the components u_i of a vector in the space of the linear representations are transformed by this series of rotations, then these components will satisfy a system of linear differential equations of the form

$$\frac{du}{dt} = p_1(t)R_1u + p_2(t)R_2u + p_3(t)R_3u.$$

It follows from this that if the matrices R_1, R_2, R_3 are known, the matrix S can be found by integrating a set of linear differential equations

$$\frac{du}{dt} = \alpha R_1u + \beta R_2u + \gamma R_3u$$

where α, β, and γ are three direction cosines, which form the parameters of the rotation. The expressions which give the vector u in the representation space in terms of t and of its initial value u_0, for sufficiently small values of t, form a set of linear substitutions having as elements *analytic* functions of αt, βt, γt: these substitutions correspond to the rotation through angle t about the axis (α, β, γ). The representation can be completed as described above.

75. Representations of the group of real rotations and analytic representations of the group of complex rotations

For real rotations in real Euclidean space the parameters αt, βt, γt are real. In a real pseudo-Euclidean space, we must take a linear combination with real coefficients of three matrices which correspond to real infinitesimal linearly independent rotations. Finally for complex rotations, αt, βt, γt can be taken as any three complex parameters. From this we deduce the following theorem.

THEOREM. *All linear representations of the group of real rotations give a linear representation of the group of complex rotations.*

We merely substitute in the expressions for the elements of the representation matrices, which are analytic functions of the real parameters of the real rotations, the values of the complex parameters. We say that the second representation (that of the complex rotation group) has been deduced from the first (that of the real rotation group) *by passing from real to complex*. The representations of the complex rotation group obtained in this way are said to be *analytic*; we shall show later (Sections 82–84) that there exist *non-analytic* representations.

The irreducible nature of linear representations is unaltered by passing from real to complex; if an analytic representation of the complex group is reducible, then since the real group is a sub-group, its representation will *a fortiori* also be reducible.

There is thus a one to one correspondence between:

(i) The analytic irreducible representations of the complex rotation group.
(ii) The irreducible representations of the real rotation group in real Euclidean space.
(iii) The irreducible representations of the group of real proper rotations in real pseudo-Euclidean space.

In the latter case the restriction to proper rotations is required; the set of all rotations does not form a continuous group.

76. Structure equations

There exist certain relations connecting the matrices R_1, R_2, R_3 which represent infinitesimal rotations in any linear representation whatsoever.

First consider two families of rotations each depending analytically on a single parameter and which give the identity rotation when this parameter is zero. Let $s(u)$ and $s(v)$ be the matrices which describe how these rotations act on vectors. To simplify the discussion, suppose that the first rotation is through an angle u about the axis (α, β, γ) and the other is through an angle v about the axis $(\alpha', \beta', \gamma')$. We form the rotation corresponding to the matrix

$$s = s(u)s(v)s(-u)s(-v);$$

this matrix can be expanded in powers of u and v; it reduces to 1 for $u = v = 0$; it also reduces to 1 if either u or v is zero; thus all terms after the first have uv as a factor; the principal part of $s - 1$ is thus of the form $uv\rho$ where the matrix ρ represents an infinitesimal rotation. To calculate ρ we form the product $s(u)s(v)s(-u)s(-v)$ where the expansion of each term is limited to the term of the first degree, i.e.,

$$s(u) = 1 + u\rho, \qquad s(v) = 1 + v\rho_2, \qquad s(-u) = 1 - u\rho_1, \qquad s(-v) = 1 - v\rho_2.$$

It follows that

$$s = 1 + uv(\rho_1\rho_2 - \rho_2\rho_1) + \cdots$$

This gives the following theorem:

THEOREM. *If the matrices ρ_1 and ρ_2 represent the operation of infinitesimal rotations on vectors, then the matrix $\rho_1\rho_2 - \rho_2\rho_1$ also represents an infinitesimal rotation.*

The important point to note is that if we consider any representation whatsoever of the rotation group, then if R_1 and R_2 correspond to ρ_1 and ρ_2, by the same argument by which the matrix $\rho_1\rho_2 - \rho_2\rho_1$ was obtained it follows that the matrix $R_1R_2 - R_2R_1$ corresponds to the rotation $\rho_1\rho_2 - \rho_2\rho_1$ in the representation.

In particular if R_1, R_2, R_3 are those matrices which represent the basic infinitesimal rotations in a linear representation, then the three matrices $R_2R_3 - R_3R_2$, $R_3R_1 - R_1R_3$, $R_1R_2 - R_2R_1$ are linear combinations of R_1, R_2 and R_3, and the coefficients in these linear combinations are the same for all linear representations of the rotation group.

Take, for instance, the spinor group, where

$$R_1 = -\tfrac{1}{2}iH_1, \qquad R_2 = -\tfrac{1}{2}iH_2, \qquad R_3 = -\tfrac{1}{2}iH_3,$$

we then have

$$R_1R_2 - R_2R_1 = -\tfrac{1}{4}(H_1H_2 - H_2H_1) = \tfrac{1}{2}H_1H_2 = -\tfrac{1}{2}iH_3 = R_3$$

and two similar results; this gives the

THEOREM. *In any linear representation whatsoever of the group of rotations, the matrices* R_1, R_2, R_3 *which represent rotations of unit angular velocity about the co-ordinate axes, satisfy the structural equations*

$$R_2R_3 - R_3R_2 = R_1, \qquad R_3R_1 - R_1R_3 = R_2, \qquad R_1R_2 - R_2R_1 = R_3.$$

These equations can easily be verified for the matrices which operate on vectors:

$$R_1 = \begin{pmatrix} 0 & 0 & 0 \\ 0 & 0 & -1 \\ 0 & 1 & 0 \end{pmatrix}, \quad R_2 = \begin{pmatrix} 0 & 0 & 1 \\ 0 & 0 & 0 \\ -1 & 0 & 0 \end{pmatrix}, \quad R_3 = \begin{pmatrix} 0 & -1 & 0 \\ 1 & 0 & 0 \\ 0 & 0 & 0 \end{pmatrix}.$$

This theorem has a converse, which forms a special case of the *second fundamental theorem* of group theory, but it is not needed here.

77. Irreducible representations of the rotation group

We now derive the linear representations of the rotation group. We shall substitute for the unknown matrices R_1, R_2, R_3 three more convenient linear combinations:

$$\left.\begin{aligned} A &= R_1 - iR_2, \\ B &= \tfrac{1}{2}(R_1 + iR_2), \\ C &= iR_3. \end{aligned}\right\} \tag{1}$$

The new structure equations are

$$AB - BA = C, \qquad AC - CA = A, \qquad BC - CB = -B. \tag{2}$$

We now seek all irreducible representations. Let λ be an eigenvalue of the matrix C; in the representation space there exists a vector u such that $Cu = \lambda u$; we shall say that it belongs to the eigenvalue λ. The equations (2) give

$$CAu = (\lambda - 1)Au, \qquad CBu = (\lambda + 1)Bu,$$

i.e., if the vector Au is not zero the matrix C has eigenvalue $\lambda - 1$, and if Bu is not zero the matrix C has eigenvalue $\lambda + 1$.

Choose the eigenvalue λ of C so that $\lambda + 1$ is not an eigenvalue; then Bu must be zero. From the equations

$$Cu = \lambda u, \qquad Bu = 0$$

it follows, on using (2), that

$$CAu = (\lambda - 1)Au, \qquad BAu = -\lambda u.$$

Then, by applying the same equations to Au, A^2u etc.,

$$CA^2u = (\lambda - 2)A^2u, \qquad BA^2u = (1 - 2\lambda)Au,$$
$$CA^3u = (\lambda - 3)A^3u, \qquad BA^3u = (3 - 3\lambda)A^2u,$$
$$\cdots\cdots\cdots\cdots\cdots\cdots\cdots\cdots\cdots\cdots$$
$$CA^pu = (\lambda - p)A^pu, \qquad BA^pu = p\left(\frac{p-1}{2} - \lambda\right)A^{p-1}u.$$

Since the number of linearly independent vectors in the space is limited, there is an integer p such that $A^{p+1}u$ is a linear combination of $u, Au, \ldots, A^pu$. Take the smallest value of p for which this holds and let

$$A^{p+1}u = \alpha_0 u + \alpha_1 Au + \alpha_2 A^2u + \cdots + \alpha_p A^pu.$$

On multiplying on the left by B a similar relation is obtained, viz.,

$$(p + 1)\left(\frac{p}{2} - \lambda\right)A^pu = -\alpha_1\lambda u + \alpha_2(1 - 2\lambda)Au + \cdots + \alpha_p p\left(\frac{p-1}{2} - \lambda\right)A^{p-1}u;$$

this is impossible unless all the coefficients are zero, i.e.,

$$p = 2\lambda, \qquad \alpha_1 = \alpha_2 = \cdots = \alpha_p = 0.$$

There remains $A^{p+1}u = \alpha_0 u$, but multiplication by C shows that $(\lambda - p - 1) \times A^{p+1}u = \alpha_0\lambda u$, i.e., $(p + 1)\alpha_0 = 0$, $\alpha_0 = 0$; this shows that

$$A^{p+1}u = 0, \qquad \lambda = \frac{p}{2}.$$

It follows from this that the independent vectors $u, Au, \ldots, A^pu$ are linearly transformed amongst themselves under infinitesimal rotations and hence also under finite rotations. The representation is irreducible and therefore of degree $p + 1$; it is uniquely determined by the integer p, since

$$Cu = \frac{p}{2}u, \qquad CAu = \left(\frac{p}{2} - 1\right)Au, \qquad CA^2u = \left(\frac{p}{2} - 2\right)A^2u, \ldots,$$
$$CA^pu = -\frac{p}{2}A^pu;$$

$$Bu = 0, \qquad BAu = -\frac{p}{2}u, \qquad BA^2u = (1 - p)Au, \ldots,$$
$$BA^pu = -\frac{p}{2}A^{p-1}u.$$

We thus have the following theorem:

THEOREM. *There exists at most one irreducible linear representation of given degree.*

Since we have just shown the existence of an irreducible representation for all given degrees, it follows that there is none other than those we have indicated. In the case where $p = 0$, the representation is of degree 1 with each of the matrices A, B, and C zero; in this representation $u' = u$.

78. Reducible representations

Take any linear representation. If it is not irreducible, there are vectors independent of u, Au, $A^2u, \ldots, A^pu$ which have already been considered, and for which we can always assume that $p/2$ is the largest eigenvalue of the matrix C. Suppose there is at least one vector v independent of u and belonging to this same eigenvalue $p/2$. By applying the same procedure to v as was applied to u, we deduce the existence of $p + 1$ vectors $v, Av, \ldots, A^pv$ which transform amongst themselves in an irreducible manner. This can be repeated for each new eigenvector which belongs to the eigenvalue $p/2$; thus suppose there are h sets of $p + 1$ vectors each of which transforms under an irreducible representation. The $h(p + 1)$ vectors are linearly independent. In effect they transform as the vectors of a linear representation which decomposes into h equivalent irreducible parts; and it is known (Section 34) that all linear relations between the $h(p + 1)$ vectors considered, can only be obtained by equating to zero all vectors of one of these parts, or the p vectors of the set

$$\alpha_1 u + \alpha_2 v + \cdots + \alpha_h w$$

$$\alpha_1 Au + \alpha_2 Av + \cdots + \alpha_h Aw$$

$$\cdots\cdots\cdots\cdots\cdots\cdots\cdots\cdots\cdot;$$

but, by hypothesis, there is no linear relation between the vectors $u, v, \ldots, w$ which belong to the eigenvalue $p/2$. It follows that the $h(p + 1)$ vectors in these h sets must be linearly independent. Let the space they determine be denoted by E_1.

Suppose now that the degree of the representation is greater than $h(p + 1)$ and that there exist eigenvalues of C corresponding to eigenvectors not situated in the space E_1. Let $q/2 < p/2$ be the greatest of these eigenvalues. Suppose that s is a vector such that $C_s = (q/2)s$; the vector Bs which belongs to the eigenvalue $(q/2) + 1$ must necessarily belong to E_1, and it must therefore be possible to obtain it by operating with B on some vector in E_1, say t, which belongs to the eigenvalue $q/2$. Replacing s by $s - t$, which is not in E_1, we see that $B(s - t) = 0$. With a change of notation, we can define a vector s, not in E_1, which satisfies

$$Cs = \frac{q}{2}s, \qquad Bs = 0.$$

The previous arguments can now be repeated to produce a set of $q + 1$ vectors $s, As, \ldots, A^qs$ which transform amongst themselves in an irreducible manner.

This process can be repeated as far as possible and gives eventually a space E' whose transformations under the rotation group form a completely reducible group, and which contains all vectors which belong to eigenvalues of C.

If the space E′ is the same as the space E of the given representation, this representation will be completely reducible. We shall now prove that this is the case.

79. Theorem of complete reducibility

We change the notation; take in the space E′ a basis formed by $u^{(1)}, u^{(2)}, \ldots, u^{(v)}$ that belong to the various eigenvalues of C. The eigenvalue λ_α to which $u^{(\alpha)}$ belongs is an integer or a half integer. We recall that $-\lambda^{(\alpha)}$ is also an eigenvalue to which there will belong as many independent vectors as belong to the eigenvalue λ_α, and that for $\lambda_\alpha \geqslant 1$ each vector belonging to the eigenvalue λ_α can be obtained by operating with B on a vector belonging to the eigenvalue $\lambda_\alpha - 1$.

Suppose the space E′ has its dimension v less than the dimension n of E. If we take two vectors in E as equivalent if their geometric difference is in E′, we obtain, in the space $\tilde{E} = E/E'$ of dimension $n - v$, a linear representation of the rotation group. Let μ be an eigenvalue of C in this representation such that $\mu + 1$ is not an eigenvalue. There exists in E a vector v (not in E′) such that

$$Cv = \mu v + \alpha_i u^{(i)}.$$

Consider the vector

$$w = v + \beta_i u^{(i)};$$

a simple calculation shows that

$$Cw = \mu w + [\alpha_i + \beta_i(\lambda_i - \mu)]u^{(i)},$$

which shows that:

(i) μ is an eigenvalue of C in its operation in E′, otherwise the constants β_i can be selected so as to make each of the coefficients $\alpha_i + \beta_i(\lambda_i - \mu)$ zero, and the vector w which is not contained in E′ is reproduced multiplied by a factor μ, which is contrary to the hypothesis that E′ contains vectors belonging to all eigenvalues of C;

(ii) the constants β_i can be chosen to make $Cw - \mu w$ belong to the eigenvalue μ.

Put $\mu = r/2$ and

$$Cw = \frac{r}{2} w + u$$

where u belongs to the eigenvalue $r/2$; then from (ii)

$$CBw = \left(\frac{r}{2} + 1\right)Bw + Bu.$$

Thus Bw must be in E′, because if it were in $\tilde{E}$ it would have to belong to the eigenvalue $r/2 + 1$ of C. Thus we can write $Bw = u^{(r/2+1)}$ + sum of vectors of E′ belonging to eigenvalues other than $r/2 + 1$.

Then $CBw = (r/2 + 1)u^{(r/2+1)}$ + sum of vectors of E′ belonging to eigenvalues other than $r/2 + 1$. Therefore the difference $CBw - (r/2 + 1)Bw$ equals a sum of vectors belonging to eigenvalues other than $r/2 + 1$; but this difference equals Bu which belongs to the eigenvalue $r/2 + 1$ i.e., $Bu = 0$. Finally, since $r/2 + 1 \geqslant 1$, there is a vector u' in E′ belonging to the eigenvalue $r/2$ such that $Bw = Bu'$. Now put $s = w - u'$, then s satisfies the fundamental equations

$$Cs = \frac{r}{2}s + u, \quad Bs = 0, \quad \text{and also} \quad Bu = 0. \tag{3}$$

The vector u generates a set of $r + 1$ vectors $u, Au, \ldots, A^r u$ which transform in an irreducible manner with $A^{r+1}u = 0$. Let us investigate the transformation of the set $s, As, \ldots, A^r s, A^{r+1}s$. Calculations similar to the above, using the structure equations (2) give

$$CAs = \left(\frac{r}{2} - 1\right)As + Au, \qquad CA^2s = \left(\frac{r}{2} - 2\right)A^2s + A^2u, \ldots$$

$$CA^h s = \left(\frac{r}{2} - h\right)A^h s + A^h u, \ldots$$

$$BAs = -\frac{r}{2}s - u, \qquad BA^2s = (1 - r)As - 2Au, \ldots$$

$$BA^h s = \frac{h}{2}(h - 1 - r)A^{h-1}s - hA^{h-1}u, \ldots$$

Let $A^{h+1}s$ be the first of the vectors $s, As, A^2s, \ldots$ which is linearly dependent on the preceding vectors and on $u, Au, \ldots, A^r u$:

$$A^{h+1}s = \alpha_0 s + \alpha_1 As + \cdots + \alpha_h A^h s + \beta_0 u + \beta_1 Au + \cdots + \beta r A^r u.$$

Operating on both sides with B gives a similar equation which does not involve $A^{h+1}s$, and must therefore be an identity. The coefficient of $A^h s$ is $[(h + 1)(h - r)]/2$, thus $h = r$; also, there is a term $-(r + 1)A^r u$ on the left-hand side and no term in $A^r u$ on the right-hand side. We have reached an impossible conclusion; it is thus absurd to assume that the space E′ is different from E, which gives the

THEOREM. *All linear representations of the rotation group* (analytic representations of the complex rotation group) *are completely reducible.*

80. The matrix $R_1^2 + R_2^2 + R_3^2$

In Quantum Mechanics the irreducible representations of the rotation group play an important rôle; each one corresponds to an energy level of an atom; the matrices $(h/i)R_1$, $(h/i)R_2$, $(h/i)R_3$ represent the three components of angular momentum; the square of the angular momentum is given by the matrix $-h^2(R_1^2 + R_2^2 + R_3^2)$. We shall show that this matrix is the product of a positive number and the unit matrix.

In any representation whatsoever of the rotation group, the matrix $R_1^2 + R_2^2 + R_3^2$ commutes with each of the matrices R_1, R_2, R_3, e.g.,

$$R_1(R_1^2 + R_2^2 + R_3^2) - (R_1^2 + R_2^2 + R_3^2)R_1 = R_1R_2^2 - R_2^2R_1 + R_1R_3^2 - R_3^2R_1$$
$$=(R_2R_1 + R_3)R_2 - R_2(R_1R_2 - R_3) + (R_3R_1 - R_2)R_3 - R_3(R_1R_3 + R_2) = 0.$$

If the representation is irreducible it follows from Section 32* that $R_1^2 + R_2^2 + R_3^2$ is a multiple of the unit matrix. Its product with any vector is therefore merely the same constant multiple of the vector.

On making the same substitution as before (Section 77)

$$A = R_1 - iR_2, \qquad B = \tfrac{1}{2}(R_1 + iR_2), \qquad C = iR_3,$$

it follows that

$$AB + BA - C^2 = R_1^2 + R_2^2 + R_3^2. \tag{4}$$

For the vector u belonging to the eigenvalue $p/2$ of C in the irreducible representation $\mathscr{D}_{p/2}$,

$$Cu = \frac{p}{2}u, \qquad Bu = 0, \qquad BAu = -\frac{p}{2}u$$

i.e.,

$$ABu + BAu - C^2u = -\frac{p}{2}\left(1 + \frac{p}{2}\right)u$$

thus

$$R_1^2 + R_2^2 + R_3^2 = -\frac{p}{2}\left(\frac{p}{2} + 1\right).$$

THEOREM. *In the irreducible representation* $\mathscr{D}_j$, *the matrix* $R_1^2 + R_2^2 + R_3^2$ *equals* $-j(j + 1)$.

81. Remarks

H. Casimir and B. L. van der Waerden†, and in a more simple treatment J. H. C. Whitehead‡ have made use of the matrix $R_1^2 + R_2^2 + R_3^2$, or rather of an analogous generalisation, to demonstrate the complete reducibility of linear representations of more general groups than the rotation group, namely of semi-simple groups; these include as a particular case the rotation group in a space of any dimension. Previously H. Weyl had given a transcendental proof§ of the same theorem for all *closed* or *compact* groups and therefore

* In fact the theorem quoted applies to matrices which commute with all matrices representing finite transformations of a group; but the matrix corresponding to a rotation through an angle θ about the axis which corresponds to the infinitesimal rotation given by R is

$$S = 1 + \theta R + \frac{\theta^2 R^2}{2} + \frac{\theta^3 R^3}{3!} + \cdots$$

and it is clear that if $R_1^2 + R_2^2 + R_3^2$ commutes with R it will also commute with s.

† *Math. Ann.*, **111**, 1935 p. 1–12.

‡ *Quarterly J. Math.*, **8**, 1937 p. 220–237.

§ H. Weyl, "Theorie der Darstellung kontinuierlicher halbeinfacher Gruppen durch lineare Transformationen", *Math. Zeitschr.*, **23**, 1925 p. 289 sqq.

for all groups which can be deduced from these by passing from real to complex (subject to the restriction of these latter to analytic representations, but this restriction can easily be disposed of).

To sum up, we have proved the theorem of complete reducibility, and we have found all irreducible representations of the rotation group in real Euclidean space and of the group of proper rotations in real pseudo-Euclidean space. The same results hold for the rotation group in complex Euclidean space except that the linear representations are limited to analytic ones.

III. LINEAR REPRESENTATIONS OF THE GROUP OF COMPLEX ROTATIONS

82. Statement of the problem

We still have to find all the continuous linear representations of the group of complex rotations. We know (Section 74) that on writing α, β, γ for the three complex parameters of a rotation and putting

$$\alpha = \alpha_1 + i\alpha_2, \qquad \beta = \beta_1 + i\beta_2, \qquad \gamma = \gamma_1 + i\gamma_2,$$

the elements of the matrices defining such a representation are analytic functions of the six real parameters $\alpha_1, \alpha_2, \beta_1, \beta_2, \gamma_1, \gamma_2$. By passing from real to complex, we obtain a linear group whose coefficients are analytic functions of the six complex parameters $\alpha_1, \alpha_2, \beta_1, \beta_2, \gamma_1, \gamma_2$, or equivalently of the six complex parameters $\alpha_1 + i\alpha_2$, $\beta_1 + i\beta_2$, $\gamma_1 + i\gamma_2$, $\alpha_1 - i\alpha_2$, $\beta_1 - i\beta_2$, $\gamma_1 - i\gamma_2$, that is of the six complex parameters $\alpha, \beta, \gamma, \bar{\alpha}, \bar{\beta}, \bar{\gamma}$, where we put $\bar{\alpha} = \alpha_1 - i\alpha_2, \ldots$, and *these six parameters are independent*.

It is clear that this linear group gives a representation of the group $\mathcal{G}$ obtained by considering rotations with parameters α, β, γ, and rotations with parameters $\bar{\alpha}$, $\bar{\beta}$, $\bar{\gamma}$. This group $\mathcal{G}$ is what is called the direct product of the group G of rotations (α, β, γ) and of the group G′ formed by the rotations $(\bar{\alpha}, \bar{\beta}, \bar{\gamma})$; each operation of $\mathcal{G}$ is a set $(\mathcal{R}, \overline{\mathcal{R}})$ of a rotation $\mathcal{R}$ with parameters (α, β, γ) and of a rotation $\overline{\mathcal{R}}$ with parameters $(\bar{\alpha}, \bar{\beta}, \bar{\gamma})$; the product of two operations in $\mathcal{G}$ namely $(\mathcal{R}, \overline{\mathcal{R}})$ and $(\mathcal{R}', \overline{\mathcal{R}}')$ is the operation $(\mathcal{R}\mathcal{R}', \overline{\mathcal{R}}\overline{\mathcal{R}}')$. This group $\mathcal{G}$ contains the subgroup G consisting of the operations $(\mathcal{R}, 1)$, when $\overline{\mathcal{R}}$ is the identity rotation, and the sub-group G′ consisting of the operations $(1, \overline{\mathcal{R}})$, when $\mathcal{R}$ is the identity rotation; the operations of G and G′ are obviously interchangeable and any operation of $\mathcal{G}$ can be expressed in one, and only one, way as the product of an operation of G by an operation of G′. $\mathcal{G}$ is said to be the *direct product* of G and of G′, and we write $\mathcal{G} = \mathrm{G} \times \mathrm{G}'$.

We are finally faced with the following problem:

PROBLEM. *Given two groups* G *and* G′ *and their direct product* $\mathrm{G} \times \mathrm{G}' = \mathcal{G}$, *and knowing all the analytic linear representations of each of the groups* G *and* G′, *to deduce all analytic linear representations of* $\mathcal{G}$.

We recall that a representation is analytic when the elements of the corresponding matrices are analytic functions of the complex parameters of the group.

In the case which we are considering we know all the analytic representations of the groups G and G′. (They are actually the same group, but the parameters are taken as being distinct).

83. Linear representations of the direct product of two groups

We assume that the theorem of complete reducibility holds for each of the component groups G and G′; this is true in our case. Any linear representation of $\mathscr{G}$ gives rise to a linear representation of its subgroup G; by hypothesis this latter representation can be decomposed into a certain number of irreducible representations. Consider one of these of degree r, and suppose there are $h - 1$ other representations equivalent to it. Let

$$x_1^{(\alpha)}, x_2^{(\alpha)}, \ldots, x_r^{(\alpha)} \qquad (\alpha = 1, 2, \ldots, h)$$

be the components of these h irreducible representations. Let s be an operation of G and t be an operation of G′; write $y_1^{(\alpha)}, y_2^{(\alpha)}, \ldots, y_r^{(\alpha)}$ for the transforms of the variables $x_1^{(\alpha)}, x_2^{(\alpha)}, \ldots, x_r^{(\alpha)}$ by the operation t. The operations s and t commute, $st = ts$. If to the variables $x_1^{(\alpha)}, \ldots, x_r^{(\alpha)}$ we apply the operation s followed by t, the variable $x_i^{(\alpha)}$ is transformed into $a_i^k x_k^{(\alpha)}$, then into $a_i^k y_k^{(\alpha)}$; if the operation t is applied first, $x_i^{(\alpha)}$ is transformed into $y_i^{(\alpha)}$, then it follows that the operation s acting on $y_i^{(\alpha)}$ must give $a_i^k y_k^{(\alpha)}$. In other words, under the operation of G the components $y_i^{(\alpha)}$ transform in the same way as the $x_i^{(\alpha)}$. It follows (Section 33) that

$$y_i^{(\alpha)} = b_\beta^\alpha x_i^{(\beta)}.$$

In particular, the operation st corresponds to the transformation $x_i^{(\alpha)} \to a_i^k b_\beta^\alpha x_k^{(\beta)}$; the matrix (a_i^k) is that which shows how the operation s of G transforms the components of an irreducible tensor x^α amongst themselves; the matrix (b_β^α) is that which shows how the operation t of G′ transforms the h tensors $x^{(1)}, x^{(2)}, \ldots, x^{(h)}$ amongst themselves. The matrices (a_i^k) define an irreducible representation of G,

$$x_i \to a_i^k x_k;$$

the matrices (b_β^α) define a representation of G′,

$$x^{(\alpha)} \to b_\beta^\alpha x^{(\beta)}.$$

It is obvious that the hr variables $x_i^{(\alpha)}$ are transformed linearly amongst themselves by the group $\mathscr{G}$; the linear representation of $\mathscr{G}$ under consideration thus decomposes into as many irreducible representations as there are non-equivalent irreducible parts in the induced representation of G′. It is also seen that the hr variables $x_i^{(\alpha)}$ transform like the products $x_i x^{(\alpha)}$ of the components of an irreducible representation of G by the components of a representation of G′. This latter is completely reducible; the representation $x_i x^{(\alpha)}$ is itself completely reducible, each part being the product of an irreducible representation of G by an irreducible representation of G′. This gives the

THEOREM. *Any tensor which is irreducible with respect to the direct product* G × G′ *of two groups* G *and* G′ *is equivalent to the product of a tensor irreducible with respect to* G *by a tensor irreducible with respect to* G′; *if the theorem of complete reducibility holds for* G *and* G′ *it also holds for their direct product.*

84. Applications to the group of complex rotations

In the case we are dealing with, any irreducible analytic tensor of the direct product of the two groups of complex rotations is equivalent to the product of a tensor $\mathscr{D}_{p/2}$ by a tensor $\mathscr{D}_{q/2}$; the former refers to rotations with parameters α, β, γ, the latter to rotations with (independent) parameters $\bar{\alpha}, \bar{\beta}, \bar{\gamma}$. If we return now to the group of complex rotations we have the same tensor, but here we regard $\bar{\alpha}, \bar{\beta}, \bar{\gamma}$ *as the complex conjugates of* α, β, γ. This is expressed in the theorem:

THEOREM. *If* ξ_0, ξ_1 *denotes an arbitrary spinor and* $\bar{\xi}_0, \bar{\xi}_1$ *its conjugate, any irreducible tensor of the complex rotation group is equivalent to a tensor which has as components monomials in* $\xi_0, \xi_1, \bar{\xi}_0, \bar{\xi}_1$ *of degree p in* ξ_0, ξ_1 *and q in* $\bar{\xi}_0, \bar{\xi}_1$.

The corresponding representation can be denoted by $\mathscr{D}_{p/q, q/2}$ and the generating polynomial can be taken as

$$(a\xi_0 + b\xi_1)^p(c\bar{\xi}_0 + d\bar{\xi}_1)^q$$

with four arbitrary parameters a, b, c, d. The degree of this representation is $(p + 1)(q + 1)$.

We shall come back later to these representations. Here we just outline the case $p = q = 1$ which gives a fourth degree tensor with components $\xi_0\bar{\xi}_0, \xi_0\bar{\xi}_1, \xi_1\bar{\xi}_0, \xi_1\bar{\xi}_1$; these components are connected by a quadratic relation. For $p = q$ the tensor is real, i.e., its components can be chosen in such a manner that for any complex rotation they transform under a linear substitution with real coefficients. Example: the nine products $x_i\bar{x}_j$ of the components of a vector by the components of its complex conjugate vector.

IV. SINGLE VALUED AND DOUBLE VALUED REPRESENTATIONS

85. Linear representations of the unimodular group in two variables are one-valued

Our investigation of irreducible linear representations of the rotation group in complex Euclidean space, of the rotation group in real Euclidean space, and of the group of proper rotations in real pseudo-Euclidean space has given single and double valued representations. For the latter two groups the double-valued representations are the $\mathcal{D}_{p/2}$ with odd p, for the first group they are the representations $\mathcal{D}_{p/2,q/2}$ with $p + q$ odd. All the representations we have found are actually also representations of the group associated with the transformation of spinors, but in this context they are single-valued. This is expressed in the following theorem:

THEOREM. *The three groups of unimodular linear substitutions in two variables which are* (1) *complex, or* (2) *unitary, or* (3) *real have no multivalued representation.*

In the case of unitary transformations there is an *a priori* topological reason for this result. Any second order unimodular unitary matrix is of the form

$$\begin{pmatrix} a & -b \\ \bar{b} & \bar{a} \end{pmatrix},$$

with $a\bar{a} + b\bar{b} = 1$. By writing

$$a = a_1 + ia_2, \qquad b = b_1 + ib_2$$

it is seen that the space of the unimodular unitary group is a manifold in which each point is defined by four real numbers a_1, a_2, b_1, b_2 for which the sum of squares equals 1; i.e., it is a spherical space of three dimensions (the hypersphere of unit radius in Euclidean space of four dimensions). This space is simply connected in the sense that all closed contours can be reduced to a point by continuous deformation. This can easily be seen by considering the inverse of the hypersphere in four dimensions with respect to a point of itself (stereographic projection); this inverse is a three-dimensional Euclidian space (including the point at infinity). Then it can be shown that if the unimodular group had a multivalued representation, on following the continuous variation of the representing matrix as the point in group space describes a suitable closed contour starting and finishing at some origin, the matrix would start as the unit matrix and finish as a different matrix. On continuously deforming the contour the final matrix will remain the same. But the contour can be deformed so as to reduce to one point—the origin. This gives a contradiction.

86. Linear representations of the real homographic group in one variable

The space of unimodular complex matrices is also simply connected, and as can be proved, this also explains the non-existence of multivalued representations of the unimodular complex group. But this result no longer holds in the space of real unimodular matrices; this is not simply connected; from the topological point of view it is homeomorphic to the inside of a torus, *thus we cannot give an a priori topological argument to show the non-existence of multivalued linear representations.* Also note that the linear representations of the group are also linear representations of the direct homographic group in a real variable z:

$$z' = \frac{az + b}{cz + d} \qquad (ad - bc > 0);$$

to each operation in this group correspond two real linear substitutions in two variables. The homographic group, therefore, has one-valued and two-valued representations, but no multivalued representations of order higher than two.

From this it can be deduced that there exist groups which have no *faithful* linear representations, i.e., no representations in which there is a one to one correspondence between group operations and representing matrices. We start from the equation

$$\tan x' = \frac{a \tan x + b}{c \tan x + d} \qquad (ad - bc > 0).$$

This equation has an infinite number of solutions for x'; for $x = 0$ we take a definite branch of arc tan b/d and follow the behaviour of x' as x varies, either from 0 to $+\infty$, or from 0 to $-\infty$. Since

$$\frac{dx'}{dx} = \frac{ad - bc}{(a \sin x + b \cos x)^2 + (c \sin x - d \cos x)^2}$$

it follows that dx/dx' always lies between two fixed positive numbers, and thus when x increases from 0 to $+\infty$, x' increases from arc tan b/d to $+\infty$, and when x decreases from 0 to $-\infty$, x' decreases from arc tan b/d to $-\infty$. We thus define a transformation on the real infinite straight line. The set of such transformations obviously forms a group, and *this group is continuous.* To verify this latter remark it is sufficient to show that one can pass continuously from the transformation with parameters a, b, c, d corresponding to one branch of arc tan b/d to the transformation with the same parameters corresponding to another branch of arc tan b/d, e.g., that which differs from it by π or by $-\pi$. To do this, take a, b, c, d as homogeneous co-ordinates of a point in three-dimensional space: to each homographic transformation corresponds a point situated in the positive region of space inside the ruled quadric surface $ad - bc = 0$. Draw through the point (a, b, c, d) a line which does not cut the quadric and take a definite point a_0, b_0, c_0, d_0 on this line. Now consider the homographic transformation with parameters $a + \lambda a_0$, $b + \lambda b_0, c + \lambda c_0, d + \lambda d_0$ where λ varies continuously from 0 to $+\infty$, then from $-\infty$ to 0; on following the value of arc tan $(b + \lambda b_0)/(d + \lambda d_0)$ from its original value, we arrive, by continuity, at this original value increased by π or this value increased by $-\pi$ (the actual value depends on the sense in which λ varies). Thus in the family of transformations of x under consideration, there is a continuous series of transformations of which the initial and final transformations both correspond to the given parameters a, b, c, d but with values of arc tan b/d which differ from each other by π.

This means that the continuous group of transformations of the variable x is such that to one transformation of tan x by the homographic group, there corresponds an infinity of transformations of x. Such a group cannot therefore have a faithful linear representation. Its manifold is simply connected. This result is all the more remarkable, since H. Weyl and Peter have shown that all closed (compact) groups always have a faithful linear representation.

We can find other groups which cover the homographic group a finite number of times and have no faithful linear representation, by considering, for example, the equation

$$\tan nx' = \frac{a \tan nx + b}{c \tan nx + d} \qquad (n \text{ an integer})$$

taken as an equation giving $z' = \tan x'$ as a function of $z = \tan x$. This equation is of degree n and has n solutions each of which provides a transformation of the real projective line z (i.e., z takes all finite values including ∞). All these transformations form a continuous group which covers the group of homographic transformations of a real variable n times. It has a faithful representation only if $n = 1$ or $n = 2$.

V. LINEAR REPRESENTATIONS OF THE GROUP OF ROTATIONS AND REVERSALS

87. Statement of the problem

We propose to find all linear representations of:

(i) The group of complex rotations and reversals.
(ii) The group of rotations and reversals in real Euclidean space.
(iii) The group of proper and improper rotations in real pseudo-Euclidean space.
(iv) The group of proper rotations and proper reversals in the latter space.
(v) The group of proper rotations and improper reversals in the same space.

In each case we have a group $\mathscr{G}$ formed by two continuous families G and G′ of which the former is a continuous group; we write $\mathscr{G} = G + G'$. We know, in each case, all the linear representations of G and we know that the theorem of complete reducibility holds. Finally, for each linear representation of G we know a representation of $\mathscr{G}$ which in G reduces to the given representation.

88. Case of irreducible representations which induce irreducible representations in the rotation group

We shall now consider those irreducible representations of $\mathscr{G}$ which give an irreducible representation of G; we shall show that there exists one and only one other non-equivalent representation of $\mathscr{G}$ which gives the same linear representation of G. Let T be the matrix which corresponds to the operation t of G′, let s be an infinitesimal operation of G and R the corresponding matrix, then the matrix TRT^{-1} corresponds to the operation tst^{-1} of G. If there is another representation, suppose T′ corresponds to t, the infinitesimal operation tst^{-1} is represented by $T'R'^{-1}$; thus for any matrix R in the representation of G under consideration

$$\mathrm{TRT}^{-1} = \mathrm{T'RT'}^{-1},$$

or

$$\mathrm{T}^{-1}\mathrm{T'R} = \mathrm{RT}^{-1}\mathrm{T'};$$

the matrix $\mathrm{T}^{-1}\mathrm{T'}$ commutes with all the matrices R of an irreducible representation and thus (by Section 32) it is a scalar, i.e., $\mathrm{T'} = m\mathrm{T}$; the coefficient m is independent of t, as can be seen by noting that all other operations of G′ are of the form st; consideration of the element t^{-1} which is in G′ shows that $m^2 = 1$ i.e., $m = -1$. Taking $-\mathrm{T}$ in place of T gives effectively a new linear representation of $\mathcal{G}$.

These two representations are not equivalent, since if they were, there would be a fixed matrix C such that

$$\mathrm{CSC}^{-1} = \mathrm{S}, \qquad \mathrm{CTC}^{-1} = -\mathrm{T};$$

the first equation requires C to be a scalar, which contradicts the second equation.

THEOREM. *Given an irreducible linear representation of the group* G, *either there is no representation of* $\mathcal{G}$ *inducing the given representation of* G, *or there are two non-equivalent representations.*

89. The converse case

Consider now an irreducible representation of $\mathcal{G}$ which induces a reducible representation of G. Let $x_1, x_2, \ldots, x_r$ be the components of one of the irreducible parts of the latter representation. Let t be an operation of G′ and suppose that the operation t transforms each of $x_1, x_2, \ldots, x_r$ into linear combinations of the variables of the representation which we shall denote by $y_1, y_2, \ldots, y_r$ (if the representation is multivalued we shall consider one of the substitutions which correspond to t). Let s be an infinitesimal operation of G and put $s' = tst^{-1}$ i.e., $ts = s't$. If we apply the operation $s't$ to the variable x_i we obtain $b_i^k y_k$, where b_i^j denotes the elements of the matrix R′ which operates on the x_i; on the other hand on applying the operation ts to x_i we obtain the same result as operating with the matrix R on the y_i; it follows that:

If the operation s' *transforms* x_i *into* $b_i^k x_k$, *the operation* $s = tst^{-1}$ *transforms* y_i *into* $b_i^k y_k$.

We deduce from this that the quantities $y_1, y_2, \ldots, y_r$ provide a linear representation, obviously irreducible, of G. By allowing t to vary in a continuous manner in G′, the x_i will be transformed into linear combinations of $y_1, y_2, \ldots, y_r$. The inverse t^{-1} of t is in G′; it follows that all operations of G′ transform $y_1, y_2, \ldots y_r$ into linear combinations of $x_1, x_2, \ldots, x_r$.

This presents two possible cases:

A. *The two linear representations of* G *spanned by the variables* x_i *and the variables* y_i *are equivalent.* Here we can suppose that by a suitable transformation of the y_i the resulting variables transform in the same manner as the

x_i for all operations of G. Let $x' = Ty$, $y' = Ux$ be the linear substitution corresponding to the operation t of G' and let R be the matrix which can operate either on the x_i or on the y_i, and corresponds to the infinitesimal operation s of G. The matrix corresponding to tst^{-1} will be TRT^{-1} operating on the x_i and URU^{-1} operating on the y_i; thus we have

$$TRT^{-1} = URU^{-1},$$

thus $U = mT$. But the variables $y_i + \sqrt{m}x_i$ are transformed amongst themselves by the operation of G' since we have

$$y' + \sqrt{m}x' = \sqrt{m}T(y + \sqrt{m}x)$$

and also

$$y' - \sqrt{m}x' = -\sqrt{m}T(y - \sqrt{m}x).$$

There must be identical linear relations between the x_i and the y_i otherwise the representation of $\mathscr{G}$ we are considering would not be irreducible. As the two irreducible representations of $\mathscr{G}$ corresponding to matrices $\sqrt{m}T$ and $-\sqrt{m}T$ are not equivalent, this is possible only if either all the variables $y_i + \sqrt{m}x_i$ are zero or all the variables $y_i - \sqrt{m}x_i$ are zero. In either case, in contradiction to the hypothesis, the representation of $\mathscr{G}$ we are considering induces an irreducible representation of G.

B. *The two linear representations of* G *spanned by the variables* x_i *and* y_i *are not equivalent*. This is now the only possible case which could arise. The variables x_i and y_i are linearly independent, otherwise all the variables x_i or all the variables y_i would be zero, which is absurd. They form all the variables of the representation of $\mathscr{G}$ under consideration. We shall see that all irreducible representations of $\mathscr{G}$ which induce in G a given pair of non-equivalent representations are equivalent to each other. Thus in a first representation of $\mathscr{G}$, let

$$x' = Uy, \qquad y' = Vx$$

be the linear substitutions (or one of the linear substitutions) corresponding to the element t of G', and let

$$x' = U'y, \qquad y' = V'x$$

be the corresponding substitutions in the second representation. If R is the matrix which operates on y_i to represent the infinitesimal s operation of G, then

$$URU^{-1} = U'RU'^{-1} \quad \text{which gives} \quad U' = mU;$$

in the same way $V' = nV$, where the coefficients m and n are constants. By considering the inverse operation, it follows that $n = 1/m$. But in the equations

$$x' = mUy, \qquad y' = \frac{1}{m}Vx$$

it is sufficient to replace x_i by mx_i and x'_i by mx'_i to obtain the equations of the first representation; this proves the proposition.

THEOREM. *If there is an irreducible representation of the group $\mathcal{G}$ which induces a reducible representation in the group G, this latter decomposes into two irreducible non-equivalent parts of the same degree, and any other irreducible representation of $\mathcal{G}$ which induces an equivalent reducible representation in G must be equivalent to the former.*

It follows from the above considerations that every operation of G′ transforms a definite irreducible tensor of G into another uniquely determined irreducible tensor. If this second tensor is equivalent to the first, either one of them provides components for two non-equivalent irreducible tensors of $\mathcal{G}$. If the given tensor and the transformed tensor are not equivalent, the set of the components of these two tensors provides one and only one irreducible tensor of $\mathcal{G}$.

90. Applications

In the case of the group of real rotations and reversals in Euclidean space, or of proper rotations and proper reversals in pseudo-Euclidean space, the $\mathcal{D}_{p/2}$ form the irreducible tensors of G; these have as generating polynomials $(a\xi_0 + b\xi_1)^p$. The reflection H_0 which changes ξ_0 into ξ_0 and ξ_1 into $-\xi_1$ leaves the component ξ_0^p unaltered; the transform of the tensor by G′ are thus equivalent to the tensor under consideration. The same conclusion can be drawn on considering the group of rotations in pseudo-Euclidean space, and also the group of proper rotations and improper reversals.

When we examine the group of complex rotations, the tensor generated by the polynomial

$$(a\xi_0 + b\xi_1)^p(c\bar{\xi}_0 + d\bar{\xi}_1)^q$$

is again transformed into an equivalent tensor by any reversal, since the component $\xi_0^p\bar{\xi}_0^q$ is invariant under the reflection H_0.

It follows that *in each of the groups mentioned we see that to each linear irreducible representation of the group G correspond two non-equivalent irreducible representations of $\mathcal{G}$, and that $\mathcal{G}$ has no other irreducible representations.*

For example, to the tensor $(a\xi_0 + b\xi_1)^p(c\bar{\xi}_0 + d\bar{\xi}_1)^q$ there correspond two irreducible tensors of $\mathcal{G}$; to the operation $\xi'_0 = \xi_0$, $\xi'_1 = -\xi_1$ on the spinors, there correspond two transformations of the generating polynomial

$$(a\xi_0 - b\xi_1)^p(c\bar{\xi}_0 - d\bar{\xi}_1)^q \quad \text{and} \quad -(a\xi_0 - b\xi_1)^p(c\bar{\xi}_0 - d\bar{\xi}_1)^q.$$

91. Case of the group of rotations and reversals in real pseudo-Euclidean space

Here we have a group made up of four continuous sets. By similar reasoning to the above it can be shown that *to each tensor of the group of proper rotations correspond four non-equivalent irreducible tensors of the whole group, and that these are the only irreducible tensors of this group.* If s, t, u, v are operations, one from each of the four sets, and in a representation they correspond to matrices S, T, U, V, then in the three related representations they correspond, respectively, to

$$S, \quad T, -U, -V;$$

$$S, -T, \quad U, -V;$$

$$S, -T, -U, \quad V.$$

In real Euclidean space there are two and only two distinct tensors which behave like a vector under the group of rotations, in pseudo-Euclidean space there are four such tensors which behave as a vector under the group of proper rotations.

Finally we note that, as can be shown by a simple argument, the theorem of complete reducibility is valid for the linear representations of $\mathcal{G} = G + G'$ if it is assumed to be valid for those of the group G; this is the case for all the applications we have just treated.

PART II

SPINORS IN SPACE OF N > 3 DIMENSIONS

SPINORS IN RIEMANNIAN GEOMETRY

CHAPTER V

SPINORS IN THE SPACE $E_{2\nu+1}$

We shall introduce spinors in a Euclidean space of odd dimension $n = 2\nu + 1$ by considering the totally isotropic subspaces of dimension ν (isotropic ν-planes) which contain the origin of the co-ordinate system. By a theorem in Chapter I any non-degenerate quadratic form can, by a suitable choice of co-ordinates, be taken as the fundamental form. We take as the $2\nu + 1$ co-ordinates $x^0, x^1, \ldots, x^\nu, x^{1'}, \ldots, x^{\nu'}$ such that the fundamental form is

$$F \equiv (x^0)^2 + x^1x^{1'} + x^2x^{2'} + \cdots + x^\nu x^{\nu'}.$$

The x^α ($\alpha = 0, 1, \ldots, \nu,\ 1', \ldots, \nu'$) can also be regarded as the contravariant components of a vector **x**. In the first sections of this chapter we shall assume that all quantities belong to the complex domain. All the subspaces we consider contain the origin.

I. ISOTROPIC ν-PLANES AND MATRICES ASSOCIATED WITH VECTORS

92. The 2^ν equations of an isotropic ν-plane

We have seen in Chapter I (Section 10) that any isotropic subspace (i.e., a subspace in which all the vectors are isotropic) has dimension at most ν. If the equations defining an isotropic subspace do not include an equation connecting $x^0, x^1, \ldots, x^\nu$ it would be possible to express the $x^{i'}$ components of each vector in the subspace as the same linear combinations of the x^0, $x^1, \ldots, x^\nu$ components which are arbitrary: such vectors do not satisfy $F = 0$. We shall establish the equations of an isotropic ν-plane assuming the general case where there is no linear relation between $x^1, x^2, \ldots, x^\nu$.

Let

$$\eta_0 \equiv \xi_0 x^0 + \xi_1 x^1 + \cdots + \xi_\nu x^\nu = 0 \tag{1}$$

be the only relation which exists between $x^0, x^1, \ldots, x^\nu$, where the coefficient ξ_0 is not zero. Taking this relation into account, the polynomial $\xi_0 F$ takes the form

$$\sum_i x^i(\xi_0 x^{i'} - \xi_i x^0),$$

which allows us to put

$$\eta_i \equiv \xi_0 x^{i'} - \xi_i x^0 + \sum_k \xi_{ik} x^k = 0 \qquad (\xi_{ij} = -\xi_{ji}). \tag{2}$$

Equations (1) and (2) define the ν-plane in terms of the constants ξ_i and ξ_{ij}. We shall introduce some further equations. Form the expression

$$\xi_i \eta_j - \xi_j \eta_i + \xi_{ij}\eta_0 \equiv \xi_0(\xi_i x^{j'} - \xi_j x^{i'} + \xi_{ij} x^0) + \sum_k (\xi_i \xi_{jk} - \xi_j \xi_{ik} + \xi_k \xi_{ij}) x^k$$

putting

$$\xi_0 \xi_{ijk} = \xi_i \xi_{jk} - \xi_j \xi_{ik} + \xi_k \xi_{ij},$$

$$\xi_0 \eta_{ij} = \xi_i \eta_j - \xi_j \eta_i + \xi_{ij} \eta_0;$$

we obtain a new set of equations

$$\eta_{ij} \equiv \xi_i x^{j'} - \xi_j x^{i'} + \xi_{ij} x^0 + \sum_k \xi_{ijk} x^k = 0. \tag{3}$$

The coefficients ξ_{ijk} have the property of components of a trivector (they change sign under an odd permutation of the indices).

Next form the expression

$$\xi_{ij}\eta_k + \xi_{jk}\eta_i + \xi_{ki}\eta_j \equiv \xi_0(\xi_{ij} x^{k'} + \xi_{jk} x^{i'} + \xi_{ki} x^{j'} - \xi_{ijk} x^0)$$
$$+ \sum_m (\xi_{ij}\xi_{km} + \xi_{jk}\xi_{im} + \xi_{ki}\xi_{jm}) x^m,$$

and put

$$\xi_0 \xi_{ijkh} = \xi_{ij}\xi_{kh} + \xi_{jk}\xi_{ih} + \xi_{ki}\xi_{jh},$$

$$\xi_0 \eta_{ijk} = \xi_{ij}\eta_k + \xi_{jk}\eta_i + \xi_{ki}\eta_j;$$

we obtain another set of equations

$$\eta_{ijk} \equiv \xi_{ij} x^{k'} + \xi_{jk} x^{i'} + \xi_{ki} x^{j'} - \xi_{ijk} x^0 + \sum_h \xi_{ijkh} x^h = 0; \tag{4}$$

the coefficients ξ_{ijkh} have the properties of components of a four-vector.

Proceeding in the same way, we eventually arrive at a set of 2^ν coefficients $\xi_{i_1 i_2 \ldots i_p}$ $(p = 1, 2, \ldots, \nu)$ which have the property of changing sign or being unaltered under odd or even permutations of the indices, and at a set of 2^ν linear forms $\eta_{i_1 i_2 \ldots i_p}$ which have the same property. The ξ_α, where α denotes the compound indices $0, i, ij, ijk, \ldots$, are defined by a recurrence relation starting from the given ξ_0, ξ_i, and ξ_{ij}. If p is even we have

(a) $$\xi_0 \xi_{i_1 i_2 \ldots i_p} = \sum_{k=1}^{k=p-1} (-1)^{k-1} \xi_{i_k i_p} \xi_{i_1 i_2 \ldots i_{k-1} i_{k+1} \ldots i_{p-1}};$$

if p is odd we have

$$\text{(b)}\quad \xi_0\xi_{i_1i_2\ldots i_p} = \sum_{k=1}^{k=p} (-1)^{k-1}\xi_{i_k}\xi_{i_1i_2\ldots i_{k-1}i_{k+1}\ldots i_p}.$$

As for the forms η_α, they are defined by

$$\text{(c)}\quad \eta_{i_1i_2\ldots i_p} = \sum_{k=1}^{k=p} (-1)^{p-k}\xi_{i_1i_2\ldots i_{k-1}i_{k+1}\ldots i_p}x^{i_k'} + (-1)^p\xi_{i_1i_2\ldots i_p}x^0$$
$$+ \sum_{m=1}^{m=\nu} \xi_{i_1i_2\ldots i_pm}x^m.$$

We also note that each ξ_α with an even number of indices $i_i, i_2, \ldots, i_p$ can be obtained, apart from a negative power of ξ_0, by forming the sum

$$\sum \xi_{j_1j_2}\xi_{j_3j_4}\cdots\xi_{j_{p-1}j_p},$$

where the indices $j_1, j_2, \ldots, j_{p-1}, j_p$ are the indices $i_1, i_2, \ldots, i_p$ in an arrangement which gives the permutation $(j_1j_2\ldots j_p)$ the same parity as $(i_1i_2\ldots i_p)$; pairs of terms with the same factors (apart from sign) are only counted once.

The 2^ν equations $\eta_\alpha = 0$ thus define, with 2^ν superfluous parameters, an isotropic ν-plane in the case $\xi_0 \neq 0$, provided that the ξ_α are restricted by the relations (a) and (b). We shall see below (Section 106) that if $\xi_0 = 0$ the same equations again define an isotropic ν-plane when the ξ_α are restricted by the relations (a) and (b) and by some other relations which will be determined. We shall also see that all isotropic ν-planes can be obtained in this way.

93. The matrix associated with a vector

We shall use the name "spinor" for any system of 2^ν quantities ξ_α, not necessarily restricted by the relations (a) and (b). We shall define later, by an appropriate convention (Sections 96 and 97), the effect produced on these quantities by a rotation or a reversal. We will then verify *a posteriori* that this convention is in accordance with the property of a certain class of spinors of being associated with an isotropic ν-plane.

We now consider the 2^ν forms η_α and the coefficient in each of them of the quantity ξ_β. If we arrange the 2^ν compound indices which are denoted by α and β in a definite order, then these coefficients will form the elements of a matrix of degree 2^ν; each of these, except in the cases where they are zero, is, apart perhaps from its sign, equal to one of the coordinates $x^0, x^1, \ldots, x^{\nu'}$, which we can consider as the contravariant components of a vector $\mathbf{x}$. *Thus we associate with any vector* $\mathbf{x}$ *a matrix* X *of degree* 2^ν. This matrix is uniquely determined once the order of the compound indices α has been chosen. For example, if $\nu = 2$ and the indices are arranged in the order 0, 1, 2, 12, then from (1), (2), and (3), we have

$$\mathrm{X} = \begin{pmatrix} x^0 & x^1 & x^2 & 0 \\ x^{1'} & -x^0 & 0 & x^2 \\ x^{2'} & 0 & -x^0 & -x^1 \\ 0 & x^{2'} & -x^{1'} & x^0 \end{pmatrix}. \tag{5}$$

For the special case where the vectors are the basis vectors, the matrices are:

$$H_0 = \begin{pmatrix} 1 & 0 & 0 & 0 \\ 0 & -1 & 0 & 0 \\ 0 & 0 & -1 & 0 \\ 0 & 0 & 0 & 1 \end{pmatrix}, \qquad H_1 = \begin{pmatrix} 0 & 1 & 0 & 0 \\ 0 & 0 & 0 & 0 \\ 0 & 0 & 0 & -1 \\ 0 & 0 & 0 & 0 \end{pmatrix},$$

$$H_2 = \begin{pmatrix} 0 & 0 & 1 & 0 \\ 0 & 0 & 0 & 1 \\ 0 & 0 & 0 & 0 \\ 0 & 0 & 0 & 0 \end{pmatrix}, \quad H_{1'} = \begin{pmatrix} 0 & 0 & 0 & 0 \\ 1 & 0 & 0 & 0 \\ 0 & 0 & 0 & 0 \\ 0 & 0 & -1 & 0 \end{pmatrix}, \quad H_{2'} = \begin{pmatrix} 0 & 0 & 0 & 0 \\ 0 & 0 & 0 & 0 \\ 1 & 0 & 0 & 0 \\ 0 & 1 & 0 & 0 \end{pmatrix}.$$

94. The fundamental theorem

The following is the fundamental property of the matrix X associated with a vector:

THEOREM. *The square of the matrix* X *associated with a vector is equal to the scalar square of the vector.*

In order to prove this theorem we note that the only non-zero elements a_α^β of X (α denotes the row suffix and β the column suffix) are

$$a^{i_1 i_2 \ldots i_p}_{i_1 i_2 \ldots i_p i_{p+1}} = x^{i'_{p+1}}, \qquad a^{i_1 i_2 \ldots i_p}_{i_1 i_2 \ldots i_p} = (-1)^p x^0, \qquad a^{i_1 i_2 \ldots i_p i_{p+1}}_{i_1 i_2 \ldots i_p} = x^{i_{p+1}}.$$

From this it follows that any element of X^2, for example

$$b^{j_1 j_2 \ldots j_q}_{i_1 i_2 \ldots i_p} = \sum_\alpha a^\alpha_{i_1 i_2 \ldots i_p} a^{j_1 j_2 \ldots j_q}_\alpha,$$

where the sum is over all the 2^ν compound indices, can only be different from zero if $p - q$ equals 0, ± 1, or, ± 2. In the case where $p - q = 2$ or 1, the q indices $j_1, j_2, \ldots, j_q$ must occur amongst the indices $i_1, i_2, \ldots, i_p$; in the case $q - p = 2$ or 1 the converse must hold; finally if $p = q$, there must be at least $p - 1$ indices in common in the two sets of indices $i_1, i_2, \ldots, i_p$ and $j_1, j_2, \ldots, j_p$.

If $p - q = 2$, then we need only consider elements $b^{j_1 j_2 \ldots j_q}_{j_1 j_2 \ldots j_q j_{q+1} j_{q+2}}$ and these are zero, since such an element is equal to the sum

$$a^{j_1 j_2 \ldots j_q j_{q+1}}_{j_1 j_2 \ldots j_q j_{q+1} j_{q+2}} a^{j_1 j_2 \ldots j_q}_{j_1 j_2 \ldots j_q j_{q+1}} + a^{j_1 j_2 \ldots j_q j_{q+2}}_{j_1 j_2 \ldots j_q j_{q+2} j_{q+1}} a^{j_1 j_2 \ldots j_q}_{j_1 j_2 \ldots j_q j_{q+2}}$$
$$= x^{j'_{q+2}} x^{j'_{q+1}} - x^{j'_{q+1}} x^{j'_{q+2}} = 0;$$

the same is true for $q - p = 2$. We find the same result for $q - p = 1$ (or $p - q = 1$), thus

$$b^{j_1 j_2 \ldots j_q}_{j_1 j_2 \ldots j_q j_{q+1}} = a^{j_1 j_2 \ldots j_q}_{j_1 j_2 \ldots j_q j_{q+1}} a^{j_1 j_2 \ldots j_q}_{j_1 j_2 \ldots j_q} + a^{j_1 j_2 \ldots j_q j_{q+1}}_{j_1 j_2 \ldots j_q j_{q+1}} a^{j_1 j_2 \ldots j_q}_{j_1 j_2 \ldots j_q j_{q+1}}$$
$$= (-1)^q x^{j'_{q+1}} x^0 + (-1)^{q+1} x^0 x^{j'_{q+1}} = 0.$$

Finally if $p = q$, we have to consider elements $b^{i_1 i_2 \dots i_p i_{p+2}}_{i_1 i_2 \dots i_p i_{p+1}}$ and $b^{i_1 i_2 \dots i_p}_{i_1 i_2 \dots i_p}$. The former are zero, as can be shown in a similar manner. As for the latter, we have

$$b^{i_1 i_2 \dots i_p}_{i_1 i_2 \dots i_p} = a^{i_1 i_2 \dots i_p}_{i_1 i_2 \dots i_p} a^{i_1 i_2 \dots i_p}_{i_1 i_2 \dots i_p} + \sum_k a^{i_1 \dots i_{k-1} i_{k+1} \dots i_p}_{i_1 i_2 \dots i_p} a^{i_1 i_2 \dots i_p}_{i_1 i_2 \dots i_{k-1} i_{k+1} \dots i_p}$$

$$+ \sum_n a^{i_1 i_2 \dots i_p h}_{i_1 i_2 \dots i_p} a^{i_1 i_2 \dots i_p}_{i_1 i_2 \dots i_p h}$$

$$= (x^0)^2 + \sum_k (-1)^{p-k} x^{i'_k} (-1)^{p-k} x^{i_k} + \sum_h x^{i_h} x^{i'_h}$$

$$= (x^0)^2 + \sum_m x^m x^{m'} = \mathbf{x}^2.$$

The following is an immediate consequence of the preceding theorem.

The scalar product of two vectors equals half the sum of the products of the associated matrices.

The proof is the same as in the space E_3 (Section 55, Theorem III).

It follows from this that if $A_1, A_2, \dots, A_n$ are the matrices associated with n orthogonal unit vectors, then

$$A_i^2 = 1, \qquad A_i A_j = -A_j A_i.$$

95. The matrix associated with a p-vector

A p-vector defined by the vectors $X_1, X_2, \dots, X_p$ can be represented by the matrix

$$\frac{1}{p!} \sum \pm X_{i_1} X_{i_2} \dots X_{i_p},$$

where the sum extends over all permutations of the indices $1, 2, \dots, p$, the sign being + or − according to whether the permutation is even or odd; such a matrix has as elements linear combinations of the components of the p-vector. We shall use the notation $\underset{(p)}{X}$ for the matrix associated with a p-vector.

If the p-vectors are orthogonal in pairs, the matrix associated with the p-vector equals $X_1 X_2 \dots X_p$. It can easily be shown by using this latter form that the square of the matrix $\underset{(p)}{X}$ associated with a p-vector is equal to the square of the measure m of the p-vector multiplied by $(-1)^{p(p-1)/2}$; for example if X and Y are two orthogonal vectors, then

$$(XY)^2 = XYXY = -XXYY = -X^2Y^2 = -m^2.$$

We shall see later (Section 98) that the matrices associated with two distinct p-vectors are themselves distinct, and that the matrix associated with an $(n - p)$-vector is the same as the matrix of a suitably chosen p-vector.

II. REPRESENTATIONS OF ROTATIONS AND REVERSALS BY MATRICES OF DEGREE 2^ν

96. The reflection associated with a unit vector

As in the space E_3, the effect on the vector X of a reflection in the hyperplane π normal to the unit vector A is given by the formula (see Section 58)

$$X' = -AXA. \tag{6}$$

For a p-vector, we have

$$\underset{(p)}{X'} = (-1)^p A \underset{(p)}{X} A. \tag{7}$$

We shall define *a priori* that the effect of this reflection on a spinor ξ is given by the formula

$$\xi' = A\xi; \tag{8}$$

the operation of reflection is two-valued since we may take either A or $-A$ as the unit vector normal to π.

The following results can be obtained by examining the effects of the operations corresponding to the basis vectors H_0, H_i, $H_{i'}$, on a spinor.

The operator H_0 reproduces each component ξ_α with or without a change of sign according to whether the compound index α contains an odd or an even number of simple indices (ξ_0 is taken as having an even number of indices—i.e., zero).

The operator H_i ($i = 1, 2, \dots, \nu$) replaces by zero those components of ξ_α for which the compound index α includes the simple index i, and adds this index to the ξ_α which do not already contain it; e.g., H_3 transforms ξ_{45} into ξ_{453} and ξ_{23} becomes zero.

The operator $H_{i'}$ makes zero those components ξ_α for which α does not contain an i and suppresses the index i in those for which α does contain the index i which must first be brought to the last position in the compound index α; for example $H_{3'}$ makes ξ_{45} zero, and transforms $\xi_{134} = -\xi_{143}$ into $-\xi_{14}$.

97. Representation of a rotation

A rotation is the resultant of an even number of reflections $A_1, A_2, \dots, A_{2k}$ ($k \leqslant \nu$); the effect it has on a vector, or more generally on a p-vector, is given by

$$\underset{(p)}{X'} = A_{2k}A_{2k-1}\dots A_2A_1\underset{(p)}{X}A_1A_2\dots A_{2k}. \tag{9}$$

The effect it has on a spinor ξ is given by

$$\xi' = A_{2k}A_{2k-1}\dots A_2A_1\xi. \tag{10}$$

If we write

$$A_{2k}A_{2k-1}\dots A_2A_1 = S,$$

the above formulae become

$$\underset{(p)}{X'} = S\underset{(p)}{X}S^{-1}, \qquad \xi' = S\xi. \tag{11}$$

The effect of a reversal can, in the same way, be represented by the formulae

$$\underset{(p)}{X'} = (-1)^p T\underset{(p)}{X}T^{-1}, \qquad \xi' = T\xi, \tag{12}$$

where the matrix T is the product of an odd number $\leqslant 2\nu + 1$ of matrices associated with unit vectors.

In particular a reflection in the origin results from the reflections associated with n unit orthogonal vectors $A_1, A_2, \dots, A_n$ and is represented by the matrix $A_1A_2\dots A_n$, that is, by the matrix associated with an n-vector of unit volume. To obtain this matrix we need only take for these n matrices

$$H_0, \qquad H_k + H_{k'}, \qquad i(H_k - H_{k'});$$

for, since $\mathbf{e}_k$ and $\mathbf{e}_{k'}$ are isotropic,

$$i^\nu H_0(H_{1'}H_1 - H_1H_{1'})\dots(H_{\nu'}H_\nu - H_\nu H_{\nu'}).$$

the last expression equals twice the scalar product of the vectors $\mathbf{e}_k$ and $\mathbf{e}_{k'}$, i.e., it equals $2g_{kk'} = 1$. On forming the product we obtain, apart from the sign,

$$i^\nu H_0(H_{1'}H_1 - H_1H_{1'})\dots(H_{\nu'}H_\nu - H_\nu H_{\nu'}).$$

By applying the rules given at the end of paragraph 96, it is found that $H_{1'}H_1 - H_1H_{1'}$, for example, reproduces each component ξ_α with or without change of sign according to whether the compound index α does not or does contain the index 1. It follows easily from this and the rule for H_0 that this product matrix reproduces each ξ_α multiplied by i^ν, which gives the following theorem.

THEOREM. *The n-vector of unit volume is represented by the scalar matrix i^ν; this matrix also represents the effect on a spinor of a reflection in the origin.*

98. The Clifford algebra

It was shown in Section 48 that p-vectors are irreducible with respect to the group of rotations. If we consider the matrix $\underset{(p)}{X}$ associated with an arbitrary p-vector, then the elements of this matrix, which are linear combinations of the components of the p-vector, are linearly transformed amongst themselves by a rotation; it follows that either all the elements are identically zero, which is absurd, or that the linear combinations which give the elements are linearly independent with respect to the ^nC_p components of the p-vector. A special case of this result is the following theorem.

THEOREM. *The matrices* $\underset{(p)}{X}$ *associated with distinct p-vectors are distinct.*

We can proceed further. Let us consider a matrix U of degree 2^ν with arbitrary complex elements. If we make the convention that under a rotation S the matrix U is transformed into SUS^{-1}, then the $2^{2\nu}$ elements of U can be regarded as forming a tensor with respect to the group of rotations. From this tensor of degree $2^{2\nu}$ we can obtain a scalar, a vector, a bivector, etc., up to a ν-vector: we need only take matrices associated with an arbitrary scalar, an arbitrary vector, etc. We thus obtain a total of $\nu + 1$ irreducible tensors which are not equivalent amonst themselves; the total number of components in these $\nu + 1$ tensors is

$$1 + {}^nC_1 + {}^nC_2 + \cdots + {}^nC_\nu = \tfrac{1}{2}2^n = 2^{2\nu};$$

i.e., is the same as the number of components in the whole tensor. There cannot be any linear relation between these $2^{2\nu}$ components, otherwise, by the theorem in Section 34, at least one of the irreducible tensors obtained must be identically zero, which is not the case. We have therefore demonstrated the following theorem.

THEOREM. *Any matrix of degree* 2^ν *can be regarded, in one and only one way, as the sum of a scalar, a vector, a bivector, ..., and a* ν*-vector.*

Matrices of degree 2^ν can be regarded as forming an algebra with $2^{2\nu}$ units over the field of complex numbers. Suppose we take as units the unit matrix 1, the matrices A_i associated with n unit orthogonal vectors, and their products in twos, threes, up to products of ν; then the law of multiplication is

$$A_i^2 = 1, \qquad A_iA_j = -A_jA_i \ (i \neq j);$$

we thus obtain the *Clifford algebra* of which the application to the representation of rotations is now evident*.

Note that the matrix associated with an $(n - p)$-vector ($p \leqslant \nu$) is identical with the matrix associated with a suitably chosen p-vector, namely the product of i^ν and the p-vector supplementary to the given $(n - p)$-vector. For example, on recalling that the product $A_1A_2 \ldots A_n$ is equal to i^ν,

$$\begin{aligned} A_{p+1}A_{p+2}\ldots A_n &= (-1)^{p(p-1)/2}(A_1A_2\ldots A_p)(A_1A_2\ldots A_pA_{p+1}\ldots A_n) \\ &= (-1)^{p(p-1)/2}i^\nu A_1A_2\ldots A_p. \end{aligned}$$

99. The tensor character of spinors

Formulae (11) and (12) show that spinors furnish a linear representation of the group of rotations and reversals; in fact if S and S′ are two rotations, then these two rotations applied successively to a vector X give as resultant rotation SS′;

$$X' = S'SXS^{-1}S'^{-1} = (S'S)X(S'S)^{-1};$$

* On this subject, see the article "Nombres complexes" in the *Encyclopédie des Sc. Math.*, French edition, adapted by E. Cartan from the German article by E. Study (*Encycl.* I, 5, 1908, no. 38, p. 463–465). For the application of this algebra in the space of special relativity, see A. Mercier, "Expression des équations de l'Électromagnétisme au moyen des nombres de Clifford", *Thesis*, Geneva, 1935; and G. Juvet, "Les rotations de l'espace euclidien à quatre dimensions, *etc.*" *Comment. Math. Helvet.*, **8**, 1936, p. 264–304.

when applied successively to a spinor they yield exactly the same resultant operation S′S. Naturally, the representation is two-valued; any rotation or any reversal can give either of two transforms when applied to a spinor. The matrices S are precisely the matrices of the representation given by spinors.

The ambiguity cannot be avoided, at least if we require the correspondence between rotations and reversals applied to vectors and rotations and reversals applied to spinors to be continuous. Thus consider the reflection A; we can vary the unit vector A in a continuous manner so as to bring it into coincidence with the vector $-A$; this gives a continuous path from the operation $\xi \rightarrow A\xi$ to the operation $\xi \rightarrow -A\xi$; both of these correspond to the same geometric operation when applied to vectors.

100. Irreducibility of spinors

We shall show that if to any linear combination whatsoever $\Sigma a^{\alpha}\xi_{\alpha}$ of the components of a spinor different rotations are applied as often as desired, we obtain 2^{ν} linearly independent combinations. We consider simple rotations through π radians, which are obtained as the resultants of the reflections associated with pairs of orthogonal unit vectors; we can also include the effects of products of pairs of such rotations, in particular we make use of the matrices

$$i(H_k + H_{k'})(H_k - H_{k'}) = i(H_{k'}H_k - H_k H_{k'})$$

and

$$(H_i \pm H_{i'})(H_j \pm H_{j'}) \qquad (i \neq j),$$

or, what amounts to the same thing,

$$H_{i'}H_i - H_i H_{i'}, \qquad H_i H_j, \qquad H_i H_{j'}, \qquad H_{i'}H_{j'} \qquad (i \neq j).$$

For definiteness we start from a linear combination $a^{\alpha}\xi_{\alpha}$, in which the coefficient of $\xi_{12\ldots p}$ is not zero. The result of applying the operation $H_{1'}H_1 - H_1H_{1'}$ is to leave unchanged those coefficients which contain the index 1, and to change the signs of the others. By addition we can obtain a new linear combination which only involves those ξ_{α} which contain the index 1. By proceeding in this manner with the indices 2, 3, . . . , p we arrive at a linear combination in which only those ξ_{α} which contain all of the indices 1, 2, . . . p occur. By using the operations $H_{(p+1)'}H_{p+1} - H_{p+1}H_{(p+1)'}, \ldots$ we can eliminate by subtraction all the ξ_{α} which contain one of the indices $p + 1, p + 2, \ldots, \nu$. In short, we see that we can deduce from the given linear combinations all the non-zero coefficients in this latter combination. From our initial assumption we thus obtain a constant multiple of $\xi_{12\ldots p}$. On applying $H_{p'}H_{(p+1)}$ to this the result is $\xi_{12\ldots(p-1)(p+1)}$, and in a similar manner we can obtain all ξ_{α} with p indices. The application of $H_{(p-1)'}H_{p'}$ enables us to obtain all components with $(p - 2)$ indices, and by using $H_{p+1}H_p$ all components with $(p + 2)$ indices. By means of similar operations we can obtain all the ξ_{α} with either an even or an odd number of indices depending upon whether p is even or odd.

We have not yet made use of the matrix H_0. Application of the operations $H_0H_{i'}$ and H_0H_i enables us to pass from the ξ_{α} with an odd number of indices to those with an even number and *vice versa.* This proves the irreducibility of spinors with respect to the group of rotations, and therefore with respect to the group of rotations and reversals. The irreducibility with respect to the latter group could obviously be proved without using the matrix H_0.

III. THE FUNDAMENTAL POLAR IN THE SPACE OF SPINORS; p-VECTORS DEFINED BY A PAIR OF SPINORS

101. The matrix C

Consider the operation defined by the matrix

$$C = (H_1 - H_{1'})(H_2 - H_{2'})\dots(H_\nu - H_{\nu'}).$$

The transform of the component $\xi_{12\dots p}$ of a spinor by the operation $H_1 - H_{1'}$ is $(-1)^p\xi_{23\dots p}$; applying the operation $H_2 - H_{2'}$ to the latter the transform is $(-1)^{p+(p-1)}\xi_{3\dots p}$, and so on up to $H_p - H_{p'}$ which gives $(-1)^{p(p+1)/2}\xi_0$; the subsequent operations give $(-1)^{p(p+1)/2}\xi_{(p+1)(p+2)\dots\nu}$. Starting from $\xi_{i_1 i_2\dots i_p}$ it is found that the transform by C is

$$(-1)^{p(p+1)/2}\xi_{i_{p+1}i_{p+2}\dots i_\nu}$$

where we assume the permutation $(i_1 i_2 \dots i_\nu)$ to be even. The only non-zero elements of C are thus the elements

$$C^{i_{p+1}\dots i_\nu}_{i_1 i_2\dots i_p} = (-1)^{p(p+1)/2}$$

where the permutation $(i_1 i_2 \dots i_\nu)$ is even.

When $\nu = 3$, for example, the rows and columns are arranged in the order

$$0,\quad 1,\quad 2,\quad 3,\quad 23,\quad 31,\quad 12,\quad 123,$$

and the matrix C is

$$C = \begin{matrix} 0 & 0 & 0 & 0 & 0 & 0 & 0 & 1 \\ 0 & 0 & 0 & 0 & -1 & 0 & 0 & 0 \\ 0 & 0 & 0 & 0 & 0 & -1 & 0 & 0 \\ 0 & 0 & 0 & 0 & 0 & 0 & -1 & 0 \\ 0 & -1 & 0 & 0 & 0 & 0 & 0 & 0 \\ 0 & 0 & -1 & 0 & 0 & 0 & 0 & 0 \\ 0 & 0 & 0 & -1 & 0 & 0 & 0 & 0 \\ 1 & 0 & 0 & 0 & 0 & 0 & 0 & 0 \end{matrix}$$

The fundamental property of the matrix C is as follows:

THEOREM. *If X is any vector whatsoever, then*

$$CX = (-1)^\nu X^T C. \tag{13}$$

It is only necessary to verify this for the basis vectors H_α; since H_0 anti-commutes with H_i and $H_{i'}$,

$$CH_0 = (-1)^\nu H_0 C = (-1)^\nu H_0^T C,$$

then

$$CH_i = (-1)^{\nu-i+1}(H_1 - H_{1'})\ldots(H_{i-1} - H_{(i-1)'})H_{i'}H_i$$
$$(H_{i+1} - H_{(i+1)'})\ldots(H_\nu - H_{\nu'}),$$
$$H_{i'}C = (-1)^{i-1}(H_1 - H_{1'})\ldots(H_{i-1} - H_{(i-1)'})H_{i'}H_i$$
$$(H_{i+1} - H_{(i+1)'})\ldots(H_\nu - H_{\nu'})$$

and since $H_{i'} = H_i^T$, the theorem is proved.

If in place of a vector X, we consider a p-vector $\underset{(p)}{X}$, we obtain

$$C\underset{(p)}{X} = (-1)^{\nu p+[p(p-1)/2]}\underset{(p)}{X}^T C. \tag{14}$$

Thus suppose, as is always permissible, that

$$\underset{(p)}{X} = X_1X_2\ldots X_p,$$

where the vectors $X_1, X_2, \ldots, X_p$ are orthogonal, then

$$C\underset{(p)}{X} = (-1)^{\nu p}X_1^TX_2^T\ldots X_p^TC = (-1)^{\nu p+[p(p-1)/2]}X_p^TX_{p-1}^T\ldots X_1^TC$$

as required.

Finally, we note the following two properties,

$$CC^T = 1, \qquad C^2 = (-1)^{\nu(\nu+1)/2}. \tag{15}$$

The first follows from

$$(H_i - H_{i'})(H_i - H_{i'})^T = (H_i - H_{i'})(H_{i'} - H_i) = H_iH_{i'} + H_{i'}H_i = 1$$

since the scalar product of the two vectors H_i and $H_{i'}$ equals $\frac{1}{2}$. The second follows from the fact that C is associated with a ν-vector formed from ν orthogonal vectors each of which has scalar square equal to -1; C^2 is thus equal to

$$(-1)^{\nu(\nu-1)/2}(-1)^\nu = (-1)^{\nu(\nu+1)/2}.$$

102. The fundamental polar form

Consider two spinors ξ, ξ' and the form $\xi^TC\xi'$, where ξ^T denotes a matrix with one row and 2^ν columns and ξ' a matrix with one column and 2^ν rows. This quantity remains unchanged when ξ and ξ' undergo the same rotation, thus under a reflection A, the form becomes

$$\xi^TA^TCA\xi',$$

but by (13)

$$A^TC = (-1)^\nu CA \quad \text{and therefore} \quad A^TCA = (-1)^\nu C;$$

the quantity under consideration is reproduced multiplied by $(-1)^\nu$. If v is even the given form is invariant under rotations and reversals, if v is odd it changes sign under a reversal. In the latter case it constitutes a tensor equivalent to an n-vector.

The relation

$$\xi^T C \xi' = 0$$

is bilinear with respect to the components of the two spinors; it defines a polar-hyperplane, "polar" in the sense that it is symmetric with respect to the two spinors; thus the left-hand side being a scalar equals its transpose; this gives

$$\xi^T C \xi' = \xi'^T C^T \xi = (-1)^{\nu(\nu+1)/2} \xi'^T C \xi.$$

This shows that the relation under consideration is symmetric with respect to the two spinors.

If $(-1)^{\nu(\nu+1)/2} = 1$, the fundamental polar is of the first type (i.e., polar with respect to a quadric); it can be derived from the quadratic form $\xi^T C \xi$; if $(-1)^{\nu(\nu+1)/2} = -1$ it is of the second type (i.e., polar with respect to a linear complex); it comes from the skew-symmetric form $[\xi^T C \xi]$.

For $\nu = 1$, we have

$$\xi^T C \xi' = \xi_0 \xi'_1 - \xi_1 \xi'_0 = [\xi_0 \xi_1];$$

for $\nu = 2$, we have

$$\xi^T C \xi' = \xi_0 \xi'_{12} - \xi_1 \xi'_2 + \xi_2 \xi'_1 - \xi_{12} \xi'_0 = [\xi_0 \xi_{12}] - [\xi_1 \xi_2];$$

for $\nu = 3$, we have

$$\tfrac{1}{2}\xi^T C \xi = \xi_0 \xi_{123} - \xi_1 \xi_{23} - \xi_2 \xi_{31} - \xi_3 \xi_{12};$$

finally for $= 4$, we have

$$\tfrac{1}{2}\xi^T C \xi = \xi_0 \xi_{1234} - \xi_1 \xi_{234} + \xi_2 \xi_{134} - \xi_3 \xi_{124} + \xi_4 \xi_{123} - \xi_{12} \xi_{34} - \xi_{23} \xi_{14} - \xi_{31} \xi_{24}.$$

We shall see later that the fundamental polar is the only polar in spinor space which is invariant under the group of rotations.

103. Reduction of the tensor $\xi_\alpha \xi'_\beta$

To summarise, we have just seen that the bilinear form $\xi^T C \xi'$ provides a tensor with a single component with respect to the group of rotations and reversals; this tensor is a scalar if ν is even, and is equivalent to an n-vector if ν is odd. More generally we shall consider the tensor $\xi_\alpha \xi'_\beta$, formed from two spinors ξ and ξ'. This tensor has $2^{2\nu}$ components. It is not irreducible; we shall show that it is completely reducible and decompose it into its irreducible parts.

To do this we form the quantity

$$\xi^T C \underset{(p)}{X} \xi',$$

where the p-vector $\underset{(p)}{X}$ is an indeterminate ($p \leqslant \nu$). Under the application of the reflection A to the spinors ξ and ξ' and to the p-vector $\underset{(p)}{X}$, this quantity becomes, on using equations (7), (8), and (14),

$$(-1)^p \xi^T A^T C A \underset{(p)}{X} \xi' = (-1)^{p+\nu} \xi^T C \underset{(p)}{X} \xi';$$

i.e., it is reproduced multiplied by $(-1)^{\nu+p} = (-1)^{\nu-p}$. It is therefore a scalar if $\nu - p$ is even, or a tensor equivalent to an n-vector if $\nu - p$ is odd.

In the first case, the quantity can be expanded in terms of the contravariant components $x^{\alpha_1 \alpha_2 \ldots \alpha_p}$ of the p-vector $\underset{(p)}{X}$ in the form

$$x^{\alpha_1 \alpha_2 \ldots \alpha_p} y_{\alpha_1 \alpha_2 \ldots \alpha_p}$$

where the $y_{\alpha_1 \alpha_2 \ldots \alpha_p}$ are bilinear with respect to ξ and ξ'. It follows from a fundamental theorem in tensor-calculus (Section 27), that the $y_{\alpha_1 \alpha_2 \ldots \alpha_p}$ constitute a tensor which is equivalent to a p-vector, and that the $y_{\alpha_1 \alpha_2 \ldots \alpha_p}$ are its covariant components.

In the second case the coefficients of the $x^{\alpha_1 \alpha_2 \ldots \alpha_p}$ are the components of an $(n - p)$-vector.

Note that in the first case p has the same parity as ν, in the second case $n - p$ has the same parity as ν.

The above shows that it is possible to obtain $\nu + 1$ irreducible tensors from the tensor $\xi_\alpha \alpha'_\beta$. The total number of components in these $\nu + 1$ tensors is, on putting p successively equal to 0, 1, 2, . . . , ν,

$$^nC_0 + {}^nC_1 + {}^nC_2 + \cdots + {}^nC_v = \tfrac{1}{2}({}^nC_0 + {}^nC_1 + \cdots + {}^nC_{n-1} + {}^nC_n)$$
$$= 2^{n-1} = 2^{2\nu};$$

it is thus the same as the total number of components of the tensor $\xi_\alpha \xi'_\beta$. We also know by a general theorem (Section 34) that the existence of an identical linear relation between the components of the $v + 1$ irreducible tensors must imply that all the components of at least one of these tensors are zero; but this cannot be so, since none of the tensors found above is identically zero; the quantity $\xi^T C \underset{(p)}{X} \xi'$ cannot be zero for all values of the two spinors ξ and ξ' and the p-vector $\underset{(p)}{X}$.

From the above considerations, it follows that the $2^{2\nu}$ components of the $\nu + 1$ irreducible tensors are linearly independent, and that the tensor $\xi_\alpha \xi'_\beta$ has been decomposed into $\nu + 1$ irreducible parts. We shall use the notation $\mathcal{T}_p$ for these different irreducible tensors.

104. The symmetric and the anti-symmetric irreducible parts of the tensor $\xi_\alpha \xi'_\beta$

It is clear that any irreducible tensor whose components are bilinear forms in the ξ_α and the ξ'_β is either symmetric or anti-symmetric. In order to distinguish between these two possibilities we proceed as above. By using equation (14) we have

$$\xi^T C \underset{(p)}{X} \xi' = \xi'^T \underset{(p)}{X}^T C^T \xi = (-1)^{\nu p + [p(p-1)/2]} \xi'^T C^T \underset{(p)}{X} \xi$$
$$= (-1)^{[\nu(\nu+1)/2] + \nu p + [p(p-1)/2]} \xi'^T C \underset{(p)}{X} \xi.$$

But

$$(-1)^{[\nu(\nu+1)/2]+\nu p+[p(p-1)/2]} = (-1)^{(\nu-p)(\nu-p+1)/2}$$

Thus if $\nu - p \equiv 0$ or -1 (mod 4), the tensor $\mathscr{T}$ is symmetric, if $\nu - p \equiv 1$ or 2 (mod 4) it is anti-symmetric.

The symmetric tensors are thus equivalent to q-vectors, where q is congruent to ν (mod 4); the anti-symmetric tensors are equivalent to q-vectors where q is congruent to $\nu + 2$ (mod 4).

The symmetric tensors are equivalent to those which are obtained on setting $\xi'_\alpha = \xi_\alpha$; they provide a decomposition of the tensor $\xi_\alpha\xi_\beta$ which has $2^{\nu-1}(2^\nu + 1)$ components; it can be verified that the sum of the binomial coefficients ${}^{2\nu+1}C_p$, where p is equal to ν plus or minus a multiple of 4, is indeed equal to $2^{\nu-1}(2^\nu + 1)$.

105. The ν-vector associated with a spinor

One of these tensors is particularly important, namely the tensor $\mathscr{T}_\nu$; it is equivalent to a ν-vector. We shall calculate its components for $\nu = 2$.

To do this we consider the expression $\xi^T C \underset{(2)}{X} \xi$; the components of the bivector are

$$\left.\begin{array}{ll}
x_{01} = \xi^T CH_0H_1\xi = -2\xi_1\xi_{12}, & x_{02} = \xi^T CH_0H_2\xi = -2\xi_1\xi_{12} \\
x_{01'} = \xi^T CH_0H_{1'}\xi = -2\xi_0\xi_2, & x_{02'} = \xi^T CH_0H_{2'}\xi = 2\xi_0\xi_1 \\
\multicolumn{2}{c}{x_{11'} = \tfrac{1}{2}\xi^T C(H_1H_{1'} - H_{1'}H_1)\xi = -\xi_0\xi_{12} - \xi_1\xi_2} \\
\multicolumn{2}{c}{x_{22'} = \tfrac{1}{2}\xi^T C(H_2H_{2'} - H_{2'}H_2)\xi = -\xi_0\xi_{12} + \xi_1\xi_2} \\
x_{12} = \xi^T CH_1H_2\xi = -\xi_{12}^2 & x_{1'2'} = \xi^T CH_{1'}H_{2'}\xi = -\xi_0^2 \\
x_{12'} = \xi^T CH_1H_{2'}\xi = \xi_1^2 & x_{1'2} = \xi^T CH_1H_2\xi = \xi_2^2
\end{array}\right\} \quad (16)$$

We have already seen at the beginning of this chapter that for $\nu = 2$ any spinor allows us to define, at least if $\xi_0 \neq 0$, an isotropic linear manifold of dimension two (biplane). We have now a method of associating a bivector with any spinor. It is easy to see that this bivector lies in this isotropic manifold. For the present we shall verify this only for those spinors in which all components except ξ_0 are zero; in this case the isotropic biplane which is associated with it has as equations

$$x^0 = x^{1'} = x^{2'} = 0.$$

Any bivector in this biplane has all its contravariant components zero except x^{12}, or what amounts to the same thing, all its covariant components zero except $x_{1'2'}$. But the bivector $\mathscr{T}_2$ determined by the spinor under consideration has just this property since its only non-zero covariant component is $x_{1'2'} = -\xi_0^2$.

This result is quite general and holds for all values of ν as we shall now show.

IV. PURE SPINORS AND THEIR INTERPRETATION AS POLARISED ISOTROPIC ν-VECTORS

At the beginning of this chapter (Section 92), in considering isotropic ν-planes, we were led to write down a system of 2^ν linear equations in n variables $x^0, x^i, x^{i'}$, the coefficients being the 2^ν components of a spinor. Provided that ξ_0 is not zero, these equations define an isotropic ν-plane if there exist certain quadratic relations (a) and (b) between the components which successively define the ξ_α with more than two indices in terms of ξ_0, the ξ_i and the ξ_{ij}. The 2^ν equations in question are represented symbolically by the matrix equation $X\xi = 0$.

106. Pure spinors and isotropic ν-planes

We say that a non-zero spinor is pure if this system of equations is of rank $\nu + 1$; they then define a ν-plane which must be isotropic, since, if the vector $\mathbf{x}$ satisfies these equations, then

$$XX\xi = X^2\xi = \mathbf{x}^2\xi = 0 \quad \text{which implies that } \mathbf{x}^2 = 0.$$

Conversely any isotropic ν-plane can be defined by a pure spinor; this has been shown when the equations of the ν-plane do not involve a linear relation between $x^1, x^2, \ldots, x^\nu$; in this case the pure spinor associated with the ν-plane has the component ξ_0 not equal to zero. To prove that the result is general, it is sufficient to appeal to the following two lemmas.

LEMMA I. *The transform of a pure spinor by a rotation, or a reflection, is a pure spinor, and the ν-plane associated with this latter spinor is the transform by the rotation, or reflection, of the ν-plane associated with the former.*

Let ξ be a pure spinor and X a vector in the associated ν-plane; as a result of the reflection A these become the spinor $\xi^1 = A\xi$ and the vector $X^1 = -AXA$; then

$$X^1\xi^1 = AXA^2\xi = -AX\xi = 0;$$

and since the vectors X' generate an isotropic ν-plane the lemma follows.

LEMMA II. *It is possible, by means of a rotation or a reversal, to transform any isotropic ν-plane into any other isotropic ν-plane.*

Suppose two isotropic ν-planes have a p-plane in common. Let X be a vector of the first ν-plane which does not lie in the second one; this vector cannot be orthogonal to all vectors in the second ν-plane, since this would require the $(\nu + 1)$-plane which contains X and the second ν-plane to be isotropic which is impossible. Let X' be a vector in the second ν-plane which is not orthogonal to X; then the vector X′ − X is not isotropic, otherwise

$$(X' - X)^2 = X'^2 + X^2 - (XX' + X'X) = -(XX' + X'X) = -2\mathbf{x}\,.\,\mathbf{x}' = 0.$$

The reflection X′ − X (i.e., that in the plane perpendicular to the direction joining the ends of the vectors X and X′) leaves invariant the p-plane common to the two given ν-planes, since any vector in this p-plane is orthogonal to X′ − X; also it will transform X into X′; it thus transforms the first ν-plane into one which has a $(p + 1)$-plane in common with the second ν-plane. On repeating the same argument several times in succession we can bring the two given ν-planes into coincidence by a number $\leqslant \nu$ of reflections.

We can go further in the case we are dealing with when n is odd; the coincidence can always be made by a rotation, since the subspace of vectors orthogonal to a ν-plane is a $\nu + 1$ plane; this cannot be isotropic, there thus exists a non-isotropic vector normal to this ν-plane and therefore a reflection which leaves the ν-plane invariant.

Let us return to our proposition. Any isotropic ν-plane can be obtained by a rotation from the ν-plane $x^0 = x^{1'} = \cdots = x^{\nu'} = 0$; we carry out this rotation on a spinor whose components other than ξ_0 are zero, then by Lemma I we obtain a pure spinor which will be associated with the given isotropic ν-plane. We thus have the following general theorem.

THEOREM. *Any isotropic ν-plane can be defined in terms of a pure spinor. The set of pure spinors is left invariant by rotations and reversals.*

107. Pure spinors make all symmetric tensors, except $\mathscr{T}_\nu$, zero

We shall now give an algebraic characterisation of pure spinors. We show that all pure spinors annul the components of the various symmetric tensors $\mathscr{T}_p$ $(p \neq \nu)$ which appear in the decomposition of the tensor $\xi_\alpha \xi_\beta$.

First, this is true for pure spinors which have all components zero except for ξ_0. Thus consider a symmetric tensor $\mathscr{T}_p$ which occurs in the expression

$$\xi^T C \underset{(p)}{X} \xi;$$

if all the ξ_α are zero except ξ_0, this expression equals ξ_0^2 multiplied by the coefficient of ξ_0 in the transformation of the component ξ_0 by the matrix $C\underset{(p)}{X}$, that is, by the coefficient of ξ_0 in the transformation of the component $\xi_{12\ldots\nu}$ by the matrix $\underset{(p)}{X}$. But the transform of $\xi_{12\ldots\nu}$ by the matrix X of a vector only involves those components ξ_α with at least $\nu - 1$ indices; on multiplying by a matrix of a further vector we obtaln an expression which only contains those ξ_α with at least $\nu - 2$ indices, and so on. Since $p < \nu$ the component ξ_0 does not appear in the transformation of $\xi_{12\ldots p}$ by $\underset{(p)}{X}$.

Using this and the result (Lemma II) that any pure spinor is a transform of a spinor of the type we have considered, the tensor $\mathscr{T}_p$ will also be zero for this spinor.

We now show that the converse holds, namely, that any spinor which annuls the symmetric tensors $\mathscr{T}_p$, other than $\mathscr{T}_\nu$, is pure. By definition, pure spinors are characterised by integral algebraic relations. Consider all those relations which are quadratic. These relations are of a tensorial nature in the sense that the right-hand sides form a tensor because they obviously form an invariant subset under rotations and reversals. Now we know from the direct theorem that included amongst the quadratic relations are all those obtained by annulling the symmetric tensors $\mathcal{T}_p$ other than $\mathcal{T}_\nu$. *There cannot be any others,* since the relations in question are linear relations between the components of the tensor $\xi_\alpha\xi_\beta$ and can only be obtained (Section 34) by equating to zero one or more of the irreducible parts of this tensor, but $\mathcal{T}_\nu$ cannot occur amongst these parts, otherwise all the components $\xi_\alpha\xi_\beta$ would be zero, from which it follows that all the ξ_α would be zero, which is absurd. It follows that *the set of quadratic relations which serve to characterise pure spinors is identical with the set obtained by annulling the symmetric tensors $\mathcal{T}_p$ other than $\mathcal{T}_\nu$.*

108. Algebraic characterisation of pure spinors

Given a spinor which annuls all the above tensors: its components satisfy the relations (a) and (b) of Section 92.

$$\xi_0\xi_{123} = \xi_1\xi_{23} + \xi_2\xi_{31} + \xi_3\xi_{12}$$
$$\cdots\cdots\cdots\cdots\cdots\cdots$$

which must necessarily exist between the components of a pure spinor. As shown at the beginning of this chapter, if $\xi_0 \neq 0$ these relations are necessary and sufficient for the spinor to be pure; it follows that every spinor, with component $\xi_0 \neq 0$, which annuls the tensors $\mathcal{T}_p$ other than $\mathcal{T}_\nu$ is pure. The result still holds if $\xi_0 = 0$, for the spinor ξ has a transform for which the component $\xi'_0 \neq 0$*, which also annuls the same tensors and as a result must be pure; ξ is thus the transform of a pure spinor and must itself be pure.

THEOREM. *In order that a spinor should be pure, it is necessary and sufficient that its components annul all tensors given by the expressions*

$$\xi^{\mathrm{T}}\underset{(p)}{\mathrm{CX}}\xi \qquad [p < \nu,\ \nu - p \equiv 0 \text{ or } 3 \pmod 4].$$

In spinor space, pure spinors thus form a manifold completely defined by a set of linearly independent quadratic equations in number $\Sigma_p^{2\nu+1}\mathrm{C}_p$, that is

$$2^{\nu-1}(2^\nu + 1) - {}^{2\nu+1}\mathrm{C}_\nu.$$

For $\nu = 3$ there is one equation, for $\nu = 4$, ${}^9\mathrm{C}_0 + {}^9\mathrm{C}_1 = 10$;

for $\nu = 5$, ${}^{11}\mathrm{C}_1 + {}^{11}\mathrm{C}_2 = 66$, for $\nu = 6$, ${}^{13}\mathrm{C}_2 + {}^{13}\mathrm{C}_3 = 364$.

109. The isotropic ν-vector associated with a pure spinor

We start with a pure spinor; as for all spinors, we can, by making use of the tensor $\mathcal{T}_\nu$, associate a ν-vector with all this spinor (Section 105). We shall show

* If all transforms of a given spinor had the component ξ_0 zero, then the smallest subspace which contains all of these transforms would be invariant under the group of rotations; this subspace is of dimension at most $2^\nu - 1$; this contradicts the irreducibility of spinors.

that this ν-vector is situated in the isotropic ν-plane defined by the spinor. We need only prove this for a particular pure spinor since the proof can be extended to any pure spinor by a suitable rotation. Take the spinor with all components, other than ξ_0, equal to zero. The argument used in the proof of Lemma II shows that the only non-zero covariant component of the ν-vector corresponds to the ν-vector $\underset{(\nu)}{X}$ which is the product of the ν vectors $H_{i'}$, since the $H_{i'}$ are the only basis vectors which, when applied to a component ξ, reduce by one the number of its indices. It follows that the only non-zero covariant component of the ν-vector associated with the spinor is

$$x_{1'2'\ldots\nu'} = \xi^T C H_{1'} H_{2'} \ldots H_{\nu'} \xi = (-1)^{\nu(\nu-1)/2} \xi_0^2;$$

the ν-plane spanned by this ν-vector has as its equations

$$x^0 = x^{1'} = x^{2'} = \cdots = x^{\nu'} = 0;$$

this is the isotropic ν-plane of the spinor we are considering.

An important consequence follows from this. Any pure spinor can be defined in a definite manner as a polarised isotropic ν-vector. Given an isotropic ν-vector the components of the spinor are determined (apart from a general change of sign) by the expressions for the tensor $\mathscr{T}_\nu$; for $\nu = 2$ these are the expressions (16) already given (Section 105).

It is obvious that any spinor can be taken, in an infinity of ways, as a sum of pure spinors; this provides a geometric interpretation of these more general spinors.

It is interesting to note that the idea of a spinor can be based on that of a vector, and conversely that the notion of a vector can be deduced from that of a spinor; at least we can form from a pure spinor an isotropic ν-vector, then a general ν-vector can be defined as the sum of isotropic ν-vectors, and a vector as a common element of a family of ν-vectors which satisfy certain conditions.

110. Intersection of two isotropic ν-planes

We use the notation $[\xi]$ for the isotropic ν-plane determined by the pure spinor ξ. The intersection of two isotropic ν-planes $[\xi]$ and $[\xi']$ is an isotropic p-plane where p can vary from 0, when the two ν-planes have no direction in common, to ν when they coincide.

There exists at least one non-isotropic vector perpendicular to both of the ν-planes; thus the space of perpendiculars to these two ν-planes is given by $2\nu - p$ independent linear equations; and is therefore a $(p + 1)$-plane which must contain the p-plane common to the two ν-planes; there must then exist a direction perpendicular to the two ν-planes which does lie in their common p-plane. This direction is not isotropic, since if it were, the $(\nu + 1)$-plane defined by it and either of the isotropic ν-planes would also be isotropic and this is impossible; since it is not isotropic it cannot be in the $(2\nu - p)$-plane formed by the two isotropic ν-planes and to which it is perpendicular.

It is always possible by a suitable rotation to take the unit vector in this direction as H_0, and to take the ν-plane $[\xi']$ as that spanned by $H_1, H_2, \ldots, H_\nu$; and we may also assume that the p-plane common to the two isotropic

ν-planes is spanned by $H_1, H_2, \ldots, H_p$. Any vector in the ν-plane $[\xi]$ must be a linear combination of $H_1, H_2 \ldots, H_p, H_{1'}, H_{2'}, \ldots, H_{p'}, H_{p+1'}, \ldots, H_{\nu'}$; but since it is orthogonal to H_1, $H_{1'}$ cannot occur in this combination, similarly for $H_{2'}, \ldots H_{p'}$. It follows that we can span the ν-plane $[\xi]$ by ν vectors

$$H_1, H_2, \ldots, H_p, \qquad K_{p+1}, \ldots, K_\nu,$$

where*

$$K_{p+1} = H_{(p+1)'} + a_{11}H_{p+1} + \cdots + a_{1(\nu-p)}H_\nu,$$

$$\cdots\cdots\cdots\cdots\cdots\cdots\cdots\cdots\cdots\cdots\cdots\cdots$$

$$K_\nu = H_{\nu'} + a_{(\nu-p)}H_{p+1} + \cdots + a_{(\nu-p)(\nu-p)}H_\nu.$$

It can be seen that the scalar products of different pairs of the vectors

$$H_0, H_1, \ldots, H_p, H_{p+1}, \ldots, H_\nu; \qquad H_{1'}, H_{2'}, \ldots, H_{p'}, K_{p+1}, \ldots, K_\nu$$

are the same as those of the vectors

$$H_0, H_1, \ldots, H_p, H_{p+1}, \ldots, H_\nu; \qquad H_{1'}, H_{2'}, \ldots, H_{p'}, H_{(p+1)'}, \ldots, H_{\nu'}.$$

They thus form a frame of reference equivalent to that of the co-ordinate frame of reference; it is thus possible by a rotation or reversal to bring the two given ν-planes into coincidence with the ν-planes

$$[H_1H_2 \ldots H_pH_{p+1} \ldots H_\nu] \quad \text{and} \quad [H_1H_2 \ldots H_pH_{(p+1)'} \ldots H_{\nu'}],$$

which have as their respective sets of equations

$$x^0 = x^{1'} = x^{2'} = \cdots = x^{p'} = x^{(p+1)'} = \cdots = x^{\nu'} = 0,$$

$$x^0 = x^{1'} = x^{2'} = \cdots = x^{p'} = x^{p+1} = \cdots = x^\nu = 0.$$

The pure spinors ξ' and ξ, corresponding to these ν-planes, are that spinor with all the components except ξ_0 zero and that spinor with all the components except $\xi_{(p+1)(p+2)\ldots\nu}$ zero; to see the latter result note that the equalities

$$H_1\xi = \cdots = H_p\xi = 0, \qquad H_{(p+1)'}\xi = \cdots = H_{\nu'}\xi = 0$$

show that all those ξ_α which contain one of the indices $1, 2, \ldots, p$ are zero, as are all those which do not contain any one of the indices $p+1, p+2, \ldots, \nu$; there remains, therefore, only the component $\xi_{(p+1)(p+2)\ldots\nu}$.

111. Condition for the intersection of two isotropic ν-planes to be of dimension p

The result obtained in the preceding section enables us to find immediately the necessary and sufficient condition for the intersection of two isotropic ν-planes $[\xi]$ and $[\xi']$ to be of dimension p. Thus if this intersection is of p dimensions all of the tensors $\mathcal{T}_0, \mathcal{T}_1, \ldots, \mathcal{T}_{p-1}$ defined by the pair of spinors ξ and ξ' are zero, this need only be proved for the two ν-planes determined above. If we form the quantity $\xi'^T C \underset{(q)}{X} \xi$, where $\underset{(q)}{X}$ is an arbitrary q-vector ($q < p$), this quantity is the product of $\xi'_0\xi_{(p+1)\ldots\nu}$ by the coefficient of $\xi_{(p+1)(p+2)\ldots\nu}$ in the transform of ξ_0 by the matrix $C\underset{(q)}{X}$, or in the transform of $\xi_{12\ldots\nu}$ by the

* The coefficients a_{ij} are not arbitrary.

matrix $\underset{(q)}{X}$. But $\underset{(q)}{X}$ is a sum of products of q matrices H_α, and as the operation H_α applied to a component ξ_β reduces the number of simple indices of this component by at most one, the effect of the operation $\underset{(q)}{X}$, applied to the component $\xi_{12\ldots\nu}$, gives an expression in which the component $\xi_{(p+1)\ldots\nu}$ cannot occur. This proves the proposition.

Conversely the tensor defined by the quantity $\mathscr{T}_p$ is not zero, since on taking $\underset{(p)}{X} = H_{1'}H_{2'}\ldots H_{p'}$, the result is $\pm\xi'_0\xi_{(p+1)\ldots\nu}$; we also see that the p-vector $\mathscr{T}_p$ is in the p-plane common to the two ν-planes.

From this there follows the theorem:

THEOREM. *For the intersection of the two isotropic ν-planes $[\xi]$ and $[\xi']$ to be of dimension p it is necessary and sufficient that the tensors $\mathscr{T}_q$ (q-vector or $(n-q)$-vector) defined by the pair of spinors ξ, ξ' should be zero for $q = 0, 1, \ldots, p-1$, but the tensor $\mathscr{T}_p$ should be non-zero.*

For example, for two isotropic ν-planes to have at least one direction in common it is necessary and sufficient that

$$\xi^T C\xi' \equiv \xi_0\xi'_{12\ldots\nu} - \xi_1\xi'_{2\ldots\nu} - \cdots - \xi_{12}\xi'_{3\ldots\nu} + \cdots + (-1)^{\nu(\nu+1)/2}\xi_{12\ldots\nu}\xi'_0 = 0;$$

that is, that the two pure spinors ξ and ξ' are conjugate with respect to the fundamental polar.

We add the following note. If the tensor $\mathscr{T}_q$ is zero, then so are the tensors $\mathscr{T}_{q-1}, \ldots, \mathscr{T}_0$. We can, as is always permissible, restrict the proof to the case where one of the ν-planes is associated with a pure spinor ξ' which has all its components except ξ'_0 zero. On examining the condition that the quantity $\xi'^T C\underset{(q)}{X}\xi$ is identically zero, we find by taking successively

$$\underset{(q)}{X} = H_{i_1'}H_{i_2'}\ldots H_{i_q'},$$

$$\underset{(q)}{X} = H_0 H_{i_1'}H_{i_2'}\ldots H_{i_{q-1}'},$$

$$\underset{(q)}{X} = H_{i_1'}H_{i_2'}\ldots H_{i_{q-2}'}(H_{i_{q-1}}H_{i_{q-1}'} - H_{i_{q-1}'}H_{i_{q-1}}),$$

that all the components of ξ which have $\nu - q$ simple indices, $\nu - q + 1$ simple indices, $\nu - q + 2$ simple indices etc., are zero. It follows that $\xi^T C\underset{(q-1)}{X}\xi$ is identically zero.

V. CASE OF REAL EUCLIDEAN SPACE

We consider the group of *real* rotations and reversals. The above expression for the fundamental form can be used if we take x^0 to be real, and the co-ordinates x^i and $x^{i'}$ to be complex conjugates. Every rotation is again the product of an even number $\leqslant 2\nu$ of reflections associated with real unit vectors A; every reversal is the product of an odd number of such reflections.

112. Conjugate vectors and p-vectors

The set of basis vectors $\mathbf{e}_0, \mathbf{e}_i, \mathbf{e}_{i'}$ has the first vector real, and the others pairs of complex conjugates. We note that the corresponding matrices $H_0, H_i, H_{i'}$ have the property that the two matrices which correspond to two complex conjugate basis vectors are transposes of each other. It follows that if we consider the matrices X and Y of two complex conjugate vectors, we have

$$X = x^0 H_0 + x^i H_i + x^{i'} H_{i'}$$
$$Y = \bar{x}^0 H_0 + \bar{x}^i H_i^T + \bar{x}^i H_{i'}^T$$

i.e.,

$$Y = \bar{X}^T.$$

In particular the matrix of a real vector is Hermitian: $\bar{X} = X^T$.

If we pass from vectors to p-vectors, we see that the p-vector conjugate to $\underset{(p)}{X}$ is $(-1)^{p(p-1)/2}\underset{p}{\bar{X}}^T$. A real p-vector has thus a Hermitian matrix if $p \equiv 0$ or 1 (mod 4), skew-Hermitian if $p \equiv 2$ or 3 (mod 4).

113. Conjugate spinors

It is important to know how to recognise when two spinors may be regarded as conjugate. If they are pure spinors, they must be derived from two complex conjugate isotropic ν-vectors. If ξ is a pure spinor, and X a vector in the isotropic ν-plane it defines, then $X\xi = 0$. Let ξ' be one of the pure spinors which define the conjugate ν-plane, and X′ be the vector conjugate to X, then

$$X'\xi' \equiv \bar{X}^T\xi' = 0;$$

by applying the formula (13) (Section 101) it follows that

$$C\bar{X}^T\xi' \equiv (-1)^\nu \bar{X} C\xi' = 0;$$

since $\bar{X}\bar{\xi} = 0$, it is reasonable to put $C\xi' = m\bar{\xi}$ or $\xi' = mC\bar{\xi}$.

To justify this formula and to find the value to be assigned to m, consider the isotropic ν-vector associated with ξ; it is defined by the expression

$$\xi^T C \underset{(\nu)}{X} \xi;$$

replace the ν-vector $\underset{(\nu)}{X}$ by its conjugate, and ξ by ξ'; this must yield an expression which is the complex conjugate of the previous one. By using formula (15) of Section 101 we see that

$$(-1)^{\nu(\nu-1)/2} m^2 \bar{\xi}^T C^T C \underset{(\nu)}{\bar{X}}^T C\bar{\xi} = \bar{\xi}^T C \underset{(\nu)}{\bar{X}} \bar{\xi};$$

noting that the right hand side equals its transpose, and taking (14) into account,

$$(-1)^{\nu(\nu-1)/2} m^2 \bar{\xi}^T \underset{(\nu)}{\bar{X}}^T C\bar{\xi} = \bar{\xi}^T \underset{(\nu)}{\bar{X}}^T C^T \bar{\xi},$$

i.e.,

$$m^2 = (-1)^{\nu(\nu-1)/2} C^T C^{-1} = (-1)^{[\nu(\nu-1)/2]+[\nu(\nu+1)/2]} = (-1)^\nu.$$

We thus have the following consistent convention:

The conjugate of a spinor ξ is the spinor $i^\nu C\bar{\xi}$.

Note that the conjugate of the conjugate of ξ is the spinor $C^2\xi = (-1)^{\nu(\nu+1)/2}\xi$. The passage from a spinor to its conjugate defines an *anti-involution* in the space of spinors; it is of the first type if $\nu \equiv 0$ or -1 (mod 4), of the second type if $\nu \equiv 1$ or 2 (mod 4). If the first case there exists a real domain in the space of spinors formed by those spinors which are equal to their conjugates. For $\nu = 3$, the real spinors are those for which $\xi = -iC\bar{\xi}$; this gives the equations

$$\xi_0 = -i\bar{\xi}_{123}, \qquad \xi_1 = i\bar{\xi}_{23}, \qquad \xi_2 = i\bar{\xi}_{31}, \qquad \xi_3 = i\bar{\xi}_{12}.$$

114. The tensor $\xi_\alpha\bar{\xi}_\beta$

The spinor conjugate to a given spinor does not transform exactly as a spinor does under reversals. In fact, under a real reflection A which transforms ξ into $A\xi$ the conjugate spinor is transformed into

$$i^\nu C\bar{A}\bar{\xi} = i^\nu CA^T\bar{\xi} = (-1)^\nu A(i^\nu C\bar{\xi}).$$

If ν is even, $C\bar{\xi}$ thus transforms in exactly the same way as a spinor; but if ν is odd, this only holds for rotations, not for reversals.

A decomposition of the tensor $\xi_\alpha\bar{\xi}_\beta$ by means of p-vectors is provided by the decomposition of the tensor $\xi_\alpha\xi'_\beta$ where we replace the spinor ξ' by the conjugate of the spinor ξ. Thus, consider the expression

$$\xi'^T C \underset{(p)}{X} \xi = i^\nu \bar{\xi}^T C^T C \underset{(p)}{X} \xi = i^\nu \bar{\xi}^T \underset{(p)}{X} \xi.$$

Under a real reflection A this expression is reproduced multiplied by $(-1)^{\nu-p}.(-1)^p$; this latter factor comes from the fact that the spinor conjugate to ξ gives rise to a p-vector or an $(n-p)$-vector according to whether p is even or odd.

The tensor $\xi_\alpha\bar{\xi}_\beta$ thus decomposes into $\nu + 1$ *irreducible tensors* which are $(2q)$-vectors. In particular for $p = 0$, we have (on omitting the constant factor i^ν) the scalar

$$\bar{\xi}^T\xi = \sum_\alpha \xi_\alpha\bar{\xi}_\alpha.$$

It is important to note that the $(2q)$-vectors thus obtained are all real.

To show this we need only show that if the p-vector $\underset{(p)}{X}$ is real, the expression $\bar{\xi}^T\underset{(p)}{X}\xi$ is real or pure imaginary, since in this case the components of the p-vector or the $(n-p)$-vector will be, to within a factor, real quantities. Now the complex conjugate of $\bar{\xi}^T X\xi$ is

$$\xi^T \underset{(p)}{X} \bar{\xi} = \bar{\xi}^T \underset{(p)}{X}^T \xi = (-1)^{p(p-1)/2}\bar{\xi}^T \underset{(p)}{X} \xi \quad \text{as required.}$$

To obtain real components for the multivector determined by two conjugate spinors, we need only to start from the expression

$$i^{p(p-1)/2}\bar{\xi}^T \underset{(p)}{X} \xi.$$

115. Example $v = 2$

For $v = 2$ the tensor $\xi_\alpha \bar{\xi}_\beta$ decomposes into three irreducible tensors:

(i) The scalar $\xi_0\bar{\xi}_0 + \xi_1\bar{\xi}_1 + \xi_2\bar{\xi}_2 + \xi_{12}\bar{\xi}_{12}$;

(ii) The four-vector given by

$$\begin{aligned}\bar{\xi}^T X \xi &= x^0 \bar{\xi}^T H_0 \xi + x^1 \bar{\xi}^T H_1 \xi + x^2 \bar{\xi}^T H_2 \xi + x^{1'} \bar{\xi}^T H_{1'} \xi + x^{2'} \bar{\xi}^T H_{2'} \xi \\ &= x^0(\xi_0\bar{\xi}_0 - \xi_1\bar{\xi}_1 - \xi_2\bar{\xi}_2 + \xi_{12}\bar{\xi}_{12}) + x'(\xi_1\bar{\xi}_0 - \xi_{12}\bar{\xi}_2) \\ &\quad + x^2(\xi_2\bar{\xi}_0 + \xi_{12}\bar{\xi}_1) + x^{1'}(\xi_0\bar{\xi}_1 - \xi_2\bar{\xi}_{12}) + x^{2'}(\xi_0\bar{\xi}_2 + \xi_1\bar{\xi}_{12});\end{aligned}$$

its contravariant components are

$$\left.\begin{aligned} x^{11'22'} &= \xi_0\bar{\xi}_0 - \xi_1\bar{\xi}_1 - \xi_2\bar{\xi}_2 + \xi_{12}\bar{\xi}_{12} \\ x^{01'22'} &= -\xi_1\bar{\xi}_0 + \xi_{12}\bar{\xi}_2 \qquad x^{0122'} = \xi_0\bar{\xi}_1 - \xi_2\bar{\xi}_{12} \\ x^{011'2'} &= -\xi_2\bar{\xi}_0 - \xi_{12}\bar{\xi}_1 \qquad x^{011'2} = \xi_0\bar{\xi}_2 + \xi_1\bar{\xi}_{12} \end{aligned}\right\} \qquad (17)$$

(iii) The bivector given by $i\bar{\xi}^T \underset{(2)}{X} \xi$ of which the ten covariant components are

$$\left.\begin{aligned} x_{01} &= i\bar{\xi}^T H_0 H_1 \xi = i(\xi_1\bar{\xi}_0 + \xi_{12}\bar{\xi}_2), \\ x_{01'} &= i\bar{\xi}^T H_0 H_{1'} \xi = -i(\xi_0\bar{\xi}_1 + \xi_2\bar{\xi}_{12}), \\ x_{02} &= i\bar{\xi}^T H_0 H_2 \xi = i(\xi_2\bar{\xi}_0 - \xi_{12}\bar{\xi}_1), \\ x_{02'} &= i\bar{\xi}^T H_0 H_{2'} \xi = -i(\xi_0\bar{\xi}_2 - \xi_1\bar{\xi}_{12}), \\ x_{12} &= i\bar{\xi}^T H_1 H_2 \xi = i\xi_{12}\bar{\xi}_0, \qquad x_{1'2'} = i\bar{\xi}^T H_{1'} H_{2'} \xi = -i\xi_0\bar{\xi}_{12}, \\ x_{12'} &= i\bar{\xi}^T H_1 H_{2'} \xi = -i\xi_1\bar{\xi}_2, \qquad x_{1'2} = i\bar{\xi}^T H_{1'} H_2 \xi = i\xi_2\bar{\xi}_1, \\ x_{11'} &= \tfrac{1}{2} i\bar{\xi}^T (H_1 H_{1'} - H_{1'} H_1)\xi = \frac{i}{2}(\xi_0\bar{\xi}_0 - \xi_1\bar{\xi}_1 + \xi_2\bar{\xi}_2 - \xi_{12}\bar{\xi}_{12}), \\ x_{22'} &= \tfrac{1}{2} i\bar{\xi}^T (H_2 H_{2'} - H_{2'} H_2)\xi = \frac{i}{2}(\xi_0\bar{\xi}_0 + \xi_1\bar{\xi}_1 - \xi_2\bar{\xi}_2 - \xi_{12}\bar{\xi}_{12}). \end{aligned}\right\} \qquad (18)$$

The measure of the four-vector and that of the bivector are both, to within a factor, equal to $\xi_0\bar{\xi}_0 + \xi_1\bar{\xi}_1 + \xi_2\bar{\xi}_2 + \xi_{12}\bar{\xi}_{12}$.

VI. CASE OF PSEUDO-EUCLIDEAN SPACE

116. The matrices I and J

We can assume that the fundamental form is reducible to a sum of $n - h$ positive squares and h negative squares, $h \leqslant v$. We can keep the expression for the fundamental form already considered by assuming that the co-ordinates

$$x^0; \qquad x^{v-h+1}, \ldots, x^v; \qquad x^{(v-h+1)'}, \ldots, x^{v'}$$

are real and that the co-ordinates x^i and $x^{i'}(i = 1, 2, \ldots, \nu - h)$ are complex conjugates. The vector

$$X = x^0H_0 + x'H_1 + \cdots + x^\nu H_\nu + x^{1'}H_{\nu'} + \cdots + x^{\nu'}H_{\nu'}$$

will then have as its conjugate the vector

$$Y = \bar{x}^0H_0 + \sum_{i=1}^{i=\nu-h} (\overline{x^i}H_{i'} + \overline{x^{i'}}H_i) + \sum_{j=\nu-h+1}^{j=\nu} (\bar{x}^jH_j + \overline{x^{j'}}H_{j'}).$$

In order to pass from X to Y we introduce the two matrices

$$I = (H_1 - H_{1'})\ldots(H_{\nu-h} - H_{(\nu-h)'}), \tag{19}$$

$$J = (H_{\nu-h+1} - H_{(\nu-h+1)'})\ldots(H_\nu \times H_{\nu'}), \tag{20}$$

which are analogues of C. We have

$$\begin{aligned} &IJ = C, \qquad II^T = JJ^T = 1 \\ &I^2 = (-1)^{[(\nu-2)(\nu-h+1)]/2}, \qquad J^2 = (-1)^{h(h+1)/2} \end{aligned} \tag{21}$$

From these results a simple calculation similar to that carried out using the matrix C leads to the following theorem.

THEOREM. *The vector* Y *conjugate to the vector* X *is given by the relation*

$$Y = (-1)^{\nu-h}I\bar{X}I^{-1} = (-1)^{\nu-h}I\bar{X}I^T = (-1)^{\nu-h}I^T\bar{X}I.$$

More generally, the p-vector $\underset{(p)}{Y}$ *conjugate to the p-vector* $\underset{(p)}{X}$ *is*

$$\underset{(p)}{Y} = (-1)^{p(\nu-h)}I\underset{(p)}{\bar{X}}I^T.$$

In the case $h = 0$ the previous results are easily obtained on noting that here $I = C$.

A vector X is real, that is, the matrix X represents a real vector if

$$\bar{X} = (-1)^{\nu-h}I^TXI = (-1)^{\nu-h}IXI^T.$$

117. Conjugate spinors

A similar argument to that given above leads us to define the conjugate of a spinor ξ as the spinor

$$\xi' = i^{\nu-h}I\bar{\xi}.$$

We verify that if in the expression

$$\xi^TC\underset{(\nu)}{X}\xi$$

the ν-vector $\underset{(\nu)}{X}$ is replaced by its conjugate, and the spinor ξ by its conjugate, the complex conjugate of the original expression is obtained. We must replace $\underset{(\nu)}{X}$ by $(-1)^{\nu(1-h)}I\underset{(\nu)}{\bar{X}}I^T$ and ξ by $i^{\nu-h}I\bar{\xi}$; the expression becomes

$$(-1)^{(\nu-1)h}\bar{\xi}^TI^TCI\underset{(\nu)}{\bar{X}}I^TI\bar{\xi} = (-1)^{(\nu-1)h}\bar{\xi}JI\underset{(\nu)}{\bar{X}}\bar{\xi} = \bar{\xi}^TC\underset{(\nu)}{\bar{X}}\bar{\xi},$$

which is certainly the complex conjugate of the original expression.

The passage from a spinor to its conjugate defines, in the space of spinors, an anti-involution which is of the first type if $I^2 = 1$, that is if $\nu - h \equiv 0$ or $-1 \pmod 4$; it is of the second type if $\nu - h \equiv 1$ or $2 \pmod 4$. In the first case there exists a real domain in the space of spinors to which belong those spinors which are equal to their conjugates.

The conjugate of a spinor transforms as a spinor under a reversal if h is of the same parity as ν.

118. The tensor product of two conjugate spinors

The product $\xi_\alpha \bar{\xi}_\beta$ of two conjugate spinors is a tensor which can be decomposed into irreducible tensors by replacing ξ' by the conjugate of ξ in the tensors $\mathscr{T}_p$. We thus have to consider the expressions

$$\bar{\xi}^T I^T \underset{(p)}{C} X \xi = \bar{\xi}^T J \underset{(p)}{X} \xi.$$

Before discussing the nature of these tensors it is important to specify under which group we consider them.

We recall (Section 12) that the group of linear substitutions which leave the fundamental form invariant splits up into four distinct connected families:

(i) *Proper rotations* which result from an even number of space-reflections and an even number of time-reflections.

(ii) *Improper rotations* which result from an odd number of space-reflections and an odd number of time-reflections.

(iii) *Proper reversals* which result from an odd number of space-reflections and even number of time-reflections.

(iv) *Improper reversals* which result from an even number of space-reflections and an odd number of time-reflections.

A space reflection is associated with a real space-like unit vector A ($A^2 = 1$); its effect on a vector X is given by the operation $-AXA$, and on a spinor ξ by the operation $A\xi$. A time reflection is associated with a real time-like unit vector A ($A^2 = -1$); its effect on a vector X is given by the operation $AXA = -AXA^{-1}$, and on a spinor ξ by the operation $iA\xi$.

The group of proper rotations and proper reversals is characterised by the property of leaving invariant the direction of time (of h dimensions).

Under a space reflection A the expression $\bar{\xi}^T J \underset{(p)}{X} \xi$ is reproduced multiplied by $(-1)^{p-h}$, whilst, under a time reflection, it is reproduced multiplied by $(-1)^{p-h+1}$.

It follows from this that under the group of proper and improper rotations and reversals the irreducible tensors into which the tensor $\xi_\alpha \bar{\xi}_\beta$ decomposes are not multivectors.

If on the other hand we restrict ourselves to the group of proper rotations and proper reversals which leave unaltered the direction of time, these tensors are equivalent to multivectors. The expression $\bar{\xi}^T J \underset{(p)}{X} \xi$ is reproduced under a proper rotation, and is multiplied by $(-1)^{p-h}$ under a proper reversal. It follows that it defines a p-vector or an $(n-p)$-vector according to whether p is of the same or opposite parity to h.

It can be shown, as in the case of a positive definite fundamental form, that the q-vectors thus generated by two conjugate spinors are real.

119. The *h*-vector generated by two conjugate spinors

A particularly interesting case is that for which $p = h$; the expression

$$\bar{\xi}^{\mathrm{T}}\mathrm{J}\underset{(h)}{\mathrm{X}}\xi$$

generates an h-vector. The time-like component of this h-vector is obtained when $\underset{(h)}{\mathrm{X}}$ is replaced by the h-vector formed from the h time-like basis vectors, namely $\mathrm{H}_{u-h+1} - \mathrm{H}_{(u-h+1)'}, \ldots, \mathrm{H}_{\nu} - \mathrm{H}_{\nu'}$: the time-like component is thus

$$\bar{\xi}^{\mathrm{T}}\mathrm{J}^2\xi = (-1)^{h(h+1)/2}\bar{\xi}^{\mathrm{T}}\xi;$$

it is, except perhaps for the sign, equal to the sum of the squares of the moduli of the 2^ν components of the spinor*.

We can deduce from this an important property, which has an application in Quantum Mechanics. If the h-vector generated by two conjugate spinors is simple, the h-plane which contains it does not contain any space-like vector. Assume that it does contain one; it is possible by a rotation to arrange that this space-like vector has all its time components zero. Let ξ' be the spinor which results from ξ under this rotation: the h-vector which results from the given h-vector under the same rotation will have its time-like component zero, since one of the vectors from which we can form it has all its time components zero; but this is impossible since this time-like component must be the sum of the squares of the moduli of the components of ξ'.

The preceding conclusion is obviously only valid if the h-vector is simple. It is interesting to carry out the calculation when $\nu = h = 2$. The components of the bivector determined by two conjugate spinors are (note that $\mathrm{J} = \mathrm{C}$);

$$\left.\begin{aligned}
x_{01} &= \bar{\xi}^{\mathrm{T}}\mathrm{CH}_0\mathrm{H}_1\xi = -(\xi_1\bar{\xi}_{12} + \xi_{12}\bar{\xi}_1),\\
x_{01'} &= \bar{\xi}^{\mathrm{T}}\mathrm{CH}_0\mathrm{H}_1\xi = -(\xi_0\bar{\xi}_2 + \xi_2\bar{\xi}_0),\\
x_{02} &= \bar{\xi}^{\mathrm{T}}\mathrm{CH}_0\mathrm{H}_2\xi = -(\xi_2\bar{\xi}_{12} + \xi_{12}\bar{\xi}_2),\\
x_{02'} &= \bar{\xi}^{\mathrm{T}}\mathrm{CH}_0\mathrm{H}_{2'}\xi = \xi_0\bar{\xi}_1 + \xi_1\bar{\xi}_0,\\
x_{12} &= \bar{\xi}^{\mathrm{T}}\mathrm{CH}_1\mathrm{H}_2\xi = -\xi_{12}\bar{\xi}_{12}, \qquad x_{1'2'} = \bar{\xi}^{\mathrm{T}}\mathrm{CH}_{1'}\mathrm{H}_{2'}\xi = \xi_2\bar{\xi}_2,\\
x_{12'} &= \bar{\xi}^{\mathrm{T}}\mathrm{CH}_1\mathrm{H}_{2'}\xi = \xi_1\bar{\xi}_1, \qquad x_{1'2} = \bar{\xi}^{\mathrm{T}}\mathrm{CH}_{1'}\mathrm{H}_2\xi = \xi_2\bar{\xi}_2,\\
x_{11'} &= \tfrac{1}{2}\bar{\xi}^{\mathrm{T}}\mathrm{C}(\mathrm{H}_1\mathrm{H}_{1'} - \mathrm{H}_{1'}\mathrm{H}_1)\xi = -\tfrac{1}{2}(\xi_0\bar{\xi}_{12} + \xi_1\bar{\xi}_2 + \xi_2\bar{\xi}_1 + \xi_{12}\bar{\xi}_0),\\
x_{22'} &= \tfrac{1}{2}\bar{\xi}^{\mathrm{T}}\mathrm{C}(\mathrm{H}_2\mathrm{H}_{2'} - \mathrm{H}_{2'}\mathrm{H}_2)\xi = -\tfrac{1}{2}(\xi_0\bar{\xi}_{12} - \xi_1\bar{\xi}_2 - \xi_2\bar{\xi}_1 + \xi_{12}\bar{\xi}_0).
\end{aligned}\right\} \quad (22)$$

The calculation shows that the bivector is simple when the scalar $\xi^{\mathrm{T}}\mathrm{C}\xi$ is zero, or, by the result of Section 111, when the conjugate isotropic biplanes $[\xi]$ and $[\xi']$ have a direction in common. It is thus true in particular if these two

* Cf. R. Brauer and H. Weyl. "Spinors in *n* dimensions, *Am. J. Math.*, **57**, 1935, p. 447.

biplanes coincide, which by the same theorem will be the case when the tensor $\mathcal{T}_1$, provided by the quantity $\xi^{\mathrm{T}}\mathrm{C}\mathrm{X}\xi$, is zero: in this case the bivector (equations 22) is nothing else but the isotropic bivector associated with a pure spinor ξ.

CHAPTER VI

SPINORS IN THE SPACE $E_{2\nu}$

I. ISOTROPIC ν-PLANES AND SEMI-SPINORS

120. Isotropic ν-planes

We can pass from a space $E_{2\nu+1}$ to a space $E_{2\nu}$ by taking $x^0 = 0$ in the former space. We thus take as fundamental form

$$F = x^1 x^{1'} + x^2 x^{2'} + \cdots + x^\nu x^{\nu'}.$$

In a space $E_{2\nu}$ isotropic manifolds are of at most ν dimensions; thus the manifold orthogonal to an isotropic p-plane is of dimension $n - p$, and since it must contain the p-plane, it is necessary that $n - p \geqslant p, p \leqslant v$. Also any isotropic ν-plane in the space $E_{2\nu}$ can be regarded as an isotropic ν-plane in the space $E_{2\nu+1}$ of which the former space is a hyperplane section, i.e., as a ν-plane orthogonal to H_0. The components of the pure spinor ξ associated with this ν-plane in the space $E_{2\nu+1}$ must thus satisfy a relation of the form

$$H_0\xi = m\xi \qquad (m \text{ is a scalar})$$

form which, on multiplying by H_0,

$$\xi = m^2\xi, \qquad m^2 = 1.$$

If $m = 1$ it follows from this that all those ξ_α with an odd number of indices are zero; if $m = -1$, all those ξ_α with an even number of indices are zero. Conversely, if a pure spinor in the space $E_{2\nu+1}$ has the property that all its components with an even (odd) number of indices are zero, then the associated ν-plane is in the space $E_{2\nu}$, since then $H_0\xi = -\xi (H_0\xi = \xi)$, which proves that the ν-plane is invariant under the reflection H_0.

121. Semi-spinors

We shall use the name "semi-spinor" in the space $E_{2\nu}$ for a system of 2^ν numbers ξ_α of which all those components with an odd (even) number of indices are zero. There are two types, semi-spinors of the first type (which we shall denote by φ) with an even number of indices, and semi-spinors of the second type (which we shall denote by ψ) with an odd number of indices. Under a reflection in the space $E_{2\nu}$, the two types of semi-spinor are interchanged: this follows from the fact that the effect of the operations H_i and $H_{i'}$ on a spinor in $E_{2\nu+1}$ is to transform each component ξ_α with an even number of indices into a component with an odd number of indices and *vice versa.* A rotation, therefore, transforms semi-spinors of each type amongst themselves. In particular the two families of isotropic ν-planes in $E_{2\nu}$ are transformed one into the other by a reversal, and transformed each into itself by a rotation.

122. Pure semi-spinors defined as polarised isotropic ν-vectors

Returning to the space $E_{2\nu+1}$, we see that a pure semi-spinor, φ for example, can be regarded as a polarised isotropic ν-vector of that space, namely, that determined by the quantity $\varphi^T C \underset{(\nu)}{X} \varphi$, where $\underset{(\nu)}{X}$ is an arbitrary ν-vector in $E_{2\nu+1}$.

It is easy to show that, in general, any p-vector in the space $E_{2\nu+1}$ can be put into the form $\underset{(p)}{X} + \underset{(p-1)}{X} H_0$, where $\underset{(p)}{X}$ is a p-vector, and $\underset{(p-1)}{X}$ a $(p-1)$-vector, both in the space $E_{2\nu}$. The quantity $\varphi^T C \underset{(\nu)}{X} \varphi$ can thus be written as (N.B. $H_0\varphi = \varphi$):

$$\varphi^T C \underset{(\nu)}{X} \varphi + \varphi^T C \underset{(\nu-1)}{X} H_0 \varphi = \varphi^T C \underset{(\nu)}{X} \varphi + \varphi^T C \underset{(\nu-1)}{X} \varphi$$

where $\underset{(\nu)}{X}$ and $\underset{(\nu-1)}{X}$ belong to $E_{2\nu}$. But we have shown, in Section 107, that the second term on the right-hand side is identically zero. The quantity $\varphi^T C \underset{(\nu)}{X} \varphi$, where the ν-vector $\underset{(\nu)}{X}$ is an arbitrary ν-vector in $E_{2\nu}$, thus defines an isotropic ν-vector in $E_{2\nu}$ of which the components are quadratic forms in the components of the pure semi-spinor φ; conversely the latter can be regarded as this polarised isotropic ν-vector.

123. Conditions for a semi-spinor to be pure

In order that a semi-spinor should be pure, it is necessary and sufficient (Section 107) that it makes each of the tensors $\varphi^T C \underset{(p)}{X} \varphi$ zero, where $p < \nu$ (a certain number of these tensors are identically zero). In this expression $\underset{(p)}{X}$ denotes an arbitrary p-vector in $E_{2\nu+1}$, but the remarks made above show that we can confine ourselves to p-vectors in $E_{2\nu}$. We know (Section 104) that the quantity is identically zero if $\nu - p \equiv 1$ or $2 \pmod 4$. Also it is identically zero if $\nu - p$ is odd; in fact the matrix $C\underset{(p)}{X}$ is the sum of products of $\nu + p$ matrices H_i and $H_{i'}$, and the effect of each of these matrices applied to a spinor is to change the parity of the number of indices of each of the components; thus if $\nu + p$ (or $\nu - p$) is odd the matrix $\underset{(p)}{\mathrm{CX}}$ changes the parity of the number of indices in the semi-spinor φ and thus makes the tensor zero. There remain for consideration only those values of p less than ν which differ from ν by a multiple of 4.

THEOREM. *In order that a semi-spinor be pure, it is necessary and sufficient that its components make identically zero all the quantities* $\xi^{\mathrm{T}}\underset{(p)}{\mathrm{CX}}\xi$, *where* $\underset{(p)}{\mathrm{X}}$ *denotes an arbitrary p-vector in* $\mathrm{E}_{2\nu}$; *p takes all those values less than* ν *which are congruent to* ν *(mod 4).*

In the $2^{\nu-1}$ dimensional space of semi-spinors of a given type, the pure semi-spinors form a set completely defined by a system of quadratic equations. A simple calculation shows that the number of linearly independent such equations is

$$^{2\nu}\mathrm{C}_{\nu-4} + {}^{2\nu}\mathrm{C}_{\nu-8} + \cdots = 2^{\nu-2}(2^{\nu-1} \times 1) - \tfrac{1}{2}{}^{2\nu}\mathrm{C}_{\nu};$$

this number is zero for $\nu = 1$, 2, and 3, one for $\nu = 4$, 10 for $\nu = 5$, 66 for $\nu = 6$, and 364 for $\nu = 7$*.

For $\nu = 4$, the relation is

$$\xi_0\xi_{1234} - \xi_{12}\xi_{34} - \xi_{23}\xi_{14} - \xi_{31}\xi_{24} = 0 \text{ (semi-spinors of the first type)}, \quad (1)$$

$$\xi_1\xi_{234} - \xi_2\xi_{134} + \xi_3\xi_{124} - \xi_4\xi_{123} = 0 \text{ (semi-spinors of the second type)}. \quad (2)$$

124. Intersection of two isotropic ν-planes

The argument of Section 110 shows that in the space $\mathrm{E}_{2\nu}$ two isotropic ν-planes $[\xi]$, $[\xi']$ can always by a rotation or a reversal be reduced to the ν-planes associated with a spinor ξ' which has all its components other than ξ_0 zero, and a spinor ξ which has all its components other than $\xi_{(p+1)(p+2)\ldots\nu}$ zero; these two ν-planes have a p dimensional intersection. We note that the semi-spinor ξ' is of the first type; and that the semi-spinor ξ is of the first type if $\nu - p$ is even, or of the second type if $\nu - p$ is odd. We thus have the

THEOREM. *The intersection of two isotropic* ν*-planes is a p-plane of which the dimension is of the same parity as* ν *if the two* ν*-planes are of the same type, or of different parity if the two* ν*-planes are of different types.*

In order that two isotropic ν*-planes of the same type,* $[\varphi]$ *and* $[\varphi']$ *for example, should have their intersection of dimension p, of the same parity as* ν, *it is necessary and sufficient that the quantities* $\varphi^{\mathrm{T}}\underset{(q)}{\mathrm{CX}}\varphi$ $(q = p - 2,$ $p - 4, \ldots)$ *be zero, and that the quantity* $\varphi^{\mathrm{T}}\underset{(p)}{\mathrm{CX}}\varphi$ *should not be zero: the p-vector determined by this latter quantity is thus isotropic and is situated in the p-plane of intersection of the given* ν*-planes.*

---

* It can easily be verified that this number equals the number of linearly independent quadratic equations which define pure spinors in the space $\mathrm{E}_{2\nu-1}$; these spinors also have $2^{\nu-1}$ components (Section 108).

In order that two isotropic ν-planes of different types, $[\varphi]$ and $[\psi]$, should have their intersection of dimension p, of different parity to ν, it is necessary and sufficient that the quantities $\varphi^{\mathrm{T}}\mathrm{C}\underset{(q)}{\mathrm{X}}\psi$ $(q = p - 2, p - 4, \ldots)$ be zero, and that the quantity $\varphi^{\mathrm{T}}\mathrm{C}\underset{(p)}{\mathrm{X}}\psi$ should not be zero: the p-vector determined by this latter quantity is then isotropic and is situated in the p-plane of intersection of the given ν-planes.

All of these results are immediate consequences of the theorem of Section 111. As in the case of the space $\mathrm{E}_{2\nu+1}$, if the quantity $\varphi^{\mathrm{T}}\mathrm{C}\underset{(p)}{\mathrm{X}}\varphi$ (p of the same parity as ν) or $\varphi^{\mathrm{T}}\mathrm{C}\underset{(p)}{\mathrm{X}}\psi$ (p of different parity from ν) vanishes identically, then the analogous quantities where p is replaced by $p - 2$, $p - 4, \ldots$ are also identically zero.

For example, for $\nu = 3$, two distinct isotropic triplanes of the same type have one direction in common, two isotropic triplanes of different types either have no direction in common or a biplane in common; the latter will be the case if

$$\xi_0\xi_{123} - \xi_1\xi_{23} - \xi_2\xi_{31} - \xi_3\xi_{12} = 0,$$

where $\xi_0, \xi_{23}, \xi_{31}, \xi_{12}$ are the components of the semi-spinor associated with the first triplane; $\xi_1, \xi_2, \xi_3, \xi_{123}$ those of the semi-spinor associated with the second.

For $\nu = 4$ two isotropic 4-planes of different types have one direction or a triplane in common; the latter will be the case if the quantity $\varphi^{\mathrm{T}}\mathrm{C}\mathrm{X}\psi$ is identically zero; this gives eight conditions of which it is sufficient to write down the two following:

$$\xi_{1234}\xi_1 - \xi_{12}\xi_{134} + \xi_{13}\xi_{124} - \xi_{14}\xi_{123} = 0,$$
$$\xi_0\xi_{234} - \xi_{23}\xi_4 + \xi_{24}\xi_3 - \xi_{34}\xi_2 = 0;$$

the ξ with an even number of indices and the ξ with an odd number of indices are respectively the components of the semi-spinors associated with the two given isotropic 4-planes. The components of each of these semi-spinors are naturally subject to the relation which characterises pure semi-spinors, namely equation (1) for the first type, equation (2) for the second type (Section 123).

II. MATRICES ASSOCIATED WITH p-VECTORS. REPRESENTATION OF ROTATIONS AND REVERSALS

125. Structure of the $\underset{(p)}{\mathrm{X}}$ matrices

The matrix $\underset{(p)}{\mathrm{X}}$ associated with a p-vector in the space $\mathrm{E}_{2\nu}$ has a special structure. We arrange the rows and columns of the matrices so that those with compound indices with an even number of simple indices come before those with compound indices with an odd number of simple indices. Since x^0 is here replaced by zero, the only non-zero elements of the matrix X associated with a vector are those for which the ordinal compound index of the row has one more, or one less, simple index than the ordinal compound index of the column. This matrix X must therefore be of the type

$$X = \begin{pmatrix} 0 & \Xi \\ H & 0 \end{pmatrix},$$

where each of the four blocks is a matrix of degree $2^{\nu-1}$. The matrix associated with a p-vector, being the sum of products of p matrices associated with vectors, will thus be of one of the forms*

$$\underset{(p)}{X} = \begin{pmatrix} \underset{(p)}{\Xi} & 0 \\ 0 & \underset{(p)}{H} \end{pmatrix} \quad \text{or} \quad \underset{(p)}{X} = \begin{pmatrix} 0 & \underset{(p)}{\Xi} \\ \underset{(p)}{H} & 0 \end{pmatrix},$$

depending upon whether p is even or odd. In particular the matrix C, which is the product of ν matrices associated with vectors, has diagonal blocks if ν is odd; the opposite holds if ν is even.

126. Effect of a reflection on a spinor

Let A be a unit vector of E_{2v}. We know, from the discussions of the preceding chapter, that the effect of the reflection associated with A on a pure semi-spinor φ can be defined by either

$$\varphi' = A\varphi \quad \text{or} \quad \varphi' = -A\varphi;$$

the same is true for a pure semi-spinor ψ. This being so, two conventions are possible, both compatible with the preceding conditions.

(i) We can associate with any reflection $A = \begin{pmatrix} 0 & \alpha \\ \beta & 0 \end{pmatrix}$ operating on a spinor $\begin{pmatrix} \varphi \\ \psi \end{pmatrix}$ the two operations

$$\varphi' = \alpha\psi, \qquad \psi' = \beta\varphi \quad \text{or} \quad \xi' = A\xi$$

$$\varphi' = -\alpha\psi, \qquad \psi' = -\beta\varphi \quad \text{or} \quad \xi' = -A\xi;$$

(ii) We can on the contrary associate with it the two operations

$$\varphi' = \alpha\psi, \qquad \psi' = -\beta\varphi \quad \text{or} \quad \xi' = AK\xi$$

$$\varphi' = -\alpha\psi, \qquad \psi' = \beta\varphi \quad \text{or} \quad \xi' = -AK\xi$$

where we write

$$K = \begin{pmatrix} -1 & 0 \\ 0 & +1 \end{pmatrix}.$$

The two conventions are equally legitimate; the first is more natural: it has the advantage of giving the same conventions in the space $E_{2\nu+1}$ and in the

* The matrix $\underset{(p)}{H}$ which will always be accompanied by the matrix $\underset{(p)}{\Xi}$, must not be confused with the matrices H_i, H'_i of the basis vectors of the space. The former are of degree $2^{\nu-1}$, the latter of degree 2^{ν}.

space $E_{2\nu}$ which is contained in it. From certain points of view the second convention introduces simplifications.

127. The two groups of rotations and reversals operating on spinors

Whichever convention we make, we have in the same set of matrices representations of rotations operating on spinors, namely $\xi' = S\xi$ where S is a product of an even number $\leqslant 2\nu$ of unit vectors.

The first convention gives for reversals the operations

$$\xi' = SA\xi,$$

A being a unit vector fixed once for all; the second convention gives, for the same reversal, the operation

$$\xi' = SAK\xi.$$

It is easy to see that the two mixed groups (of rotations and reversals) thus defined are distinct and do not have the same structure; the product of two reversals SA and SA′ is not the same as that of the two reversals SAK and SA′K; for example

$$AKA'K = -AA' \quad \text{and not} \quad AA'.$$

Considered as operating on vectors these groups are of course identical.

We shall in what follows adopt the first, more natural, convention. We merely note the property of the matrix K of anticommuting with any vector.

128. Irreducibility of spinors and semi-spinors

The arguments of Section 100 show that spinors are irreducible with respect to the group of rotations and reversals (defined in either way) and under the group of rotations decompose into two irreducible parts, which are the two types of semi-spinor.

129. Decomposition of matrices of degree 2^ν into sums of *p*-vectors

The elements of the matrix $\underset{(p)}{X}$ associated with a p-vector are linear combinations of the ${}^nC_p$ components of this p-vector and it is clear that they are not all identically zero. Since p-vectors are irreducible with respect to the group of rotations and reversals, and since the elements of $\underset{(p)}{X}$ transform linearly amongst themselves under the elements of this group, it follows that there are ${}^nC_p$ independent elements. Thus the matrices associated with two distinct $\underset{(p)}{X}$-vectors are distinct. It can be shown as in Section 98 that from an arbitrary matrix of degree 2^ν it is possible to obtain a scalar, a vector, ..., an n-vector, and that irreducible tensors are not equivalent amongst themselves under the group of rotations and reversals. None of these irreducible tensors is identically zero and as the sum of their degrees is

$$1 + {}^nC_1 + {}^nC_2 + \cdots {}^nC_n = 2^n = 2^{2\nu},$$

which equals the number of elements in the matrix under consideration, we have the

THEOREM. *Any matrix of degree 2^ν can be regarded in one and only one way as the sum of a scalar, a vector, a bivector, etc...., and an n-vector in the space of 2ν dimensions.*

Here, again, with a different interpretation, we have the Clifford Algebra.

It is interesting to see the structure of the matrix associated with an n-vector of $E_{2\nu}$: it is identical with the matrix associated in the space $E_{2\nu+1}$ with the vector perpendicular to $E_{2\nu}$, and is consequently of the form $\begin{pmatrix} m & 0 \\ 0 & -m \end{pmatrix}$.

In particular the reflection in the origin is represented by the matrix $\begin{pmatrix} i^\nu & 0 \\ 0 & -i^\nu \end{pmatrix}$ and by the matrix $\begin{pmatrix} -i^\nu & 0 \\ 0 & i^\nu \end{pmatrix}$.

130. The structure of the matrices $\underset{(\nu)}{X}$

Under a rotation the elements of each of the matrices $\underset{(p)}{\Xi}$, $\underset{(p)}{H}$ which form a matrix $\underset{(p)}{X}$ (Section 125) undergo a linear substitution. If $p \neq \nu$, the p-vector is irreducible with respect to the group of rotations; it follows that the elements of each of the matrices $\underset{(p)}{\Xi}$, $\underset{(p)}{H}$ are linearly independent combinations of the ${}^nC_p$ components of the p-vector represented by $\underset{(p)}{X}$.

The argument fails if $p = \nu$; a ν-vector decomposes under the group of rotations into two irreducible non-equivalent tensors, namely the semi-ν-vectors (Section 49); the elements of one of the matrices $\underset{(\nu)}{\Xi}$ will be *either* linearly independent combinations of the ${}^{2\nu}C_\nu$ components of the ν-vector, *or* linearly independent combinations of the $\frac{1}{2}{}^{2\nu}C_\nu$ components of one of the semi-ν-vectors into which the given ν-vector decomposes. To show that the second possibility holds, it is sufficient to show the existence of a ν-vector for which the matrix $\underset{(\nu)}{H}$ is identically zero, and of a ν-vector for which the matrix $\underset{(\nu)}{\Xi}$ is identically zero. For example, the ν-vector $H_1H_2 \ldots H_\nu$ is represented by a matrix which, when it operates on a spinor, annuls all the components except ξ_0 which is transformed into $\xi_{12\ldots\nu}$; thus the matrix has only one non-zero row, the row labelled by the suffix 0; it follows that the matrix $\underset{(\nu)}{H}$ is identically zero; the ν-vector under consideration, which is isotropic, decomposes into two semi-ν-vectors of which one is identically zero. In the same way the matrix $H_1H_2 \ldots H_{\nu-1}H_{\nu'}$ annuls all the components ξ_α of a spinor, except ξ_ν which it transforms into $(-1)^{\nu-1}\xi_{12\ldots(\nu-1)}$; the only row of this matrix which is not identically zero is the row ν; this implies that the component $\underset{(\nu)}{\Xi}$ of this matrix is identically zero.

Since any isotropic ν-sector can, by means of a rotation, be reduced to a multiple of one of the two preceding ν-vectors, it can be seen that the matrices associated with isotropic ν-vectors of the first type have the component $\underset{(\nu)}{\mathrm{H}}$ identically zero, and the matrices associated with isotropic ν-vectors of the second type have the component $\underset{(\nu)}{\Xi}$ identically zero; these ν-vectors are really particular cases of semi-ν-vectors.

To sum up, any p-vector ($p \neq \nu$) can be associated in two different ways with a matrix of degree $2^{\nu-1}$ (either $\underset{(p)}{\Xi}$ or $\underset{(p)}{\mathrm{H}}$), and any semi-$\nu$-vector can be associated with a definite matrix of degree $2^{\nu-1}$ ($\underset{(\nu)}{\Xi}$ for semi-ν-vectors of the first type, $\underset{(\nu)}{\mathrm{H}}$ for the others).

III. THE DECOMPOSITION OF THE PRODUCT OF TWO SPINORS

131. Decomposition with respect to the group of rotations and reversals

The product $\xi_\alpha \xi'_\beta$ of two spinors can be decomposed by considering the $n + 1$ quantities

$$\xi^{\mathrm{T}} \mathrm{C} \underset{(p)}{\mathrm{X}} \xi' \qquad (p = 0, 1, 2, \ldots, n).$$

Under the reflection associated with a unit vector A, this quantity is reproduced multiplied by $(-1)^{\nu-p}$; it defines, therefore, as in $\mathrm{E}_{2\nu+1}$, a p-vector or an $(n-p)$-vector, according to whether $\nu - p$ is even or odd.

From this we deduce the

THEOREM. *The product of two spinors is completely reducible with respect to the group of rotations and reversals and decomposes into a scalar, a vector, a bivector, etc., . . . , an n-vector.*

The conclusion rests on the fact that the total number, $2^{2\nu}$, of products $\xi_\alpha \xi'_\beta$ equals the sum of the degrees of the non-equivalent irreducible tensors found.

As in Section 104, it can be shown that the p-vectors obtained are symmetric with respect to the two spinors ξ and ξ' if $\nu - p \equiv 0$ or 3 (mod 4), antisymmetric if $\nu - p \equiv 1$ or 2 (mod 4). By making the two spinors the same, the symmetric tensors yield the decomposition of the tensor $\xi_\alpha \xi_\beta$. For $\nu = 2$ this gives a vector and a bivector; for $\nu = 3$ this gives a six-vector, a trivector, and a bivector.

132. Decomposition of the product of two semi-spinors with respect to the group of rotations

Let us take first the product of two semi-spinors of the same type, for example φ and φ'. The quantity $\varphi^{\mathrm{T}} \mathrm{C} \underset{(p)}{\mathrm{X}} \varphi'$, where φ is the matrix $\binom{\varphi}{0}$ with 2^ν rows and 1 column, is different from zero only if $\nu + p$ is even. Thus by giving to p values of the same and different parities to ν, we obtain tensors irreducible with respect to the group of rotations; where only the $(\nu - 2p)$-vectors are not zero.

Finally by giving p the value ν, we obtain a semi-ν-vector of the first or of the second type: if ν is even the matrix $\underset{(\nu)}{\Xi}$ operates, if ν is odd the matrix $\underset{(\nu)}{\mathrm{H}}$ operates.

The total number of components of the non-equivalent irreducible tensors thus obtained from the tensor $\varphi_\alpha \varphi'_\beta$ equals half the sum

$$\cdots + {}^n\mathrm{C}_{\nu-4} + {}^n\mathrm{C}_{\nu-2} + {}^n\mathrm{C}_{\nu} + {}^n\mathrm{C}_{\nu+2} + {}^n\mathrm{C}_{\nu+4} + \cdots = \tfrac{1}{2}2^n = 2^{2\nu-1}$$

i.e., $2^{2\nu-2}$ which is the number of products $\varphi_\alpha \varphi'_\beta$.

THEOREM. *The product of two semi-spinors of the same type decomposes, with respect to the group of rotations, into a semi-ν-vector, a $(\nu - 2)$-vector, a $(\nu - 4)$-vector, etc.*

Note that the product of two semi-spinors of the same given type is not equivalent to the product of two semi-spinors of the other type; the semi-ν-vectors of different types are not equivalent.

Let us now take the product $\varphi_\alpha \psi_\beta$ of two semi-spinors of different types. It is necessary for this to consider the quantities $\xi^{\mathrm{T}}\mathrm{C}\underset{(p)}{\mathrm{X}}\xi'$, where ξ denotes the matrix $\binom{\varphi}{0}$ and ξ' the matrix $\binom{0}{\psi}$, and where p is of different parity from ν. Thus from the given product tensor we can obtain a $(\nu - 1)$-vector, a $(\nu - 3)$-vector, etc., tensors which are not equivalent amongst themselves. The total number of their components is again $2^{2\nu-2}$, equal to the number of products $\varphi_\alpha \psi_\beta$, which gives the

THEOREM. *The product of two semi-spinors of different types decomposes, with respect to the group of rotations, into a $(\nu - 1)$-vector, a $(\nu - 3)$-vector, etc.*

133. Decomposition of the tensors $\varphi_\alpha \varphi_\beta$ and $\psi_\alpha \psi_\beta$

It is necessary to take those p-vectors in the preceding decomposition which are symmetric with respect to the two semi-spinors φ, φ' or ψ, ψ'. This gives the

THEOREM. *The tensor $\varphi_\alpha \varphi_\beta$ decomposes, with respect to the group of rotations, into a semi-ν-vector, a $(\nu - 4)$-vector, a $(\nu - 8)$-vector, etc. The same result holds for $\psi_\alpha \psi_\beta$.*

For $\nu = 2$ this gives a semi-bivector; for $\nu = 3$ a semi-trivector; for $\nu = 4$ a scalar and a semi-4-vector, etc.

134. Application to the group of rotations when ν is odd

In the case of odd ν, the quantity $\varphi^{\mathrm{T}}\mathrm{C}\psi$ constructed from two semi-spinors of different types is invariant under any rotation.

This quantity is equal to

$$\sum (-1)^{p(p+1)/2} \xi_{i_1 i_2 \ldots i_p} \xi'_{i_{p+1} \ldots i_\nu},$$

where the sum extends over all even permutations $(i_1 i_2 \ldots i_\nu)$ of the indices $1, 2, \ldots, \nu$ with the integer p taking all even values. Arrange the rows and columns in the matrix associated with any vector, first those with an even number of indices in some definite order, and then the others in a corresponding order, so that the compound index $(i_1 i_2 \ldots i_p)$ corresponds to the compound index $(-1)^{p(p+1)/2}(i_{p+1}, \ldots i_\nu)$. If we now denote the $2^{\nu-1}$ components of the semi-spinor φ by $\varphi_1, \varphi_2 \ldots$ and the corresponding components of the semi-spinor ψ by $\psi_1, \psi_2, \ldots$ we see that any rotation leaves invariant the sum $\varphi_1\psi_1 + \varphi_2\psi_2 + \ldots + \varphi_{2^{\nu-1}}\psi_{2^{\nu-1}}$. Therefore if Σ is the matrix of degree $2^{\nu-1}$ which shows how the φ_α transform under a rotation, then the components ψ_α must transform according to $(\Sigma^T)^{-1}$, thus the matrix of degree 2^ν associated with this rotation is

$$S = \begin{pmatrix} \Sigma & 0 \\ 0 & (\Sigma^T)^{-1} \end{pmatrix}.$$

If this is applied to a vector X, it will be transformed into

$$X' = SXS^{-1} = \begin{pmatrix} \Sigma & 0 \\ 0 & (\Sigma^T)^{-1} \end{pmatrix} \begin{pmatrix} 0 & \Xi \\ H & 0 \end{pmatrix} \begin{pmatrix} \Sigma^{-1} & 0 \\ 0 & \Sigma^T \end{pmatrix},$$

in particular

$$\Xi' = \Sigma \Xi \Sigma^T, \qquad H' = (\Sigma^T)^{-1} H \Sigma^{-1}.$$

We thus arrive at the following remarkable result:

THEOREM. *In the space $E_{2\nu}$, ν odd, any rotation applied to a vector transforms the associated matrix Ξ of degree $2^{\nu-1}$ into Ξ' given by*

$$\Xi' = \Sigma \Xi \Sigma^T. \tag{3}$$

In particular, taking a different point of view, we see that *knowledge of the effect produced by a rotation on a semi-spinor of the first type determines without ambiguity the effect produced on a semi-spinor of the second type.*

135. Case when ν is even

In the case where ν is even there is no comparable situation. This is because knowing one of those operations $\varphi' = \Sigma\varphi$ which correspond to the given rotation allows us to make either of two distinct corresponding operations on the semi-spinors ψ and on vectors. For example the identity operation $\varphi' = \varphi$ can correspond either to the identity rotation, which gives $\psi' = \psi$, $X' = X$, or to the reflection in the origin; this reflection corresponds to either of the two matrices

$$\begin{pmatrix} 1 & 0 \\ 0 & -1 \end{pmatrix} \quad \text{or} \quad \begin{pmatrix} -1 & 0 \\ 0 & 1 \end{pmatrix}$$

(here $i^\nu = \pm 1$); the first of these gives

$$\varphi' = \varphi, \qquad \psi' = -\psi, \qquad X' = -X.$$

In the same way, the identity operation on semi-spinors of the second type can be made to correspond to either of two distinct operations on semi-spinors of the first type.

To sum up, rotations can operate on three sorts of objects: semi-spinors of the first type, semi-spinors of the second type, and vectors; they thus provide three groups; between any pair of them there exists a two-valued correspondence in each direction if ν is even.

136. Note

The matrices Σ and T which were introduced as operators on semi-spinors describing the effect of a rotation are all unimodular. To show this let us find when the determinant of the matrix Ξ, of degree $2^{\nu-1}$, associated with a vector, can be zero. If it is zero, then it is possible to find a non-zero semi-spinor ψ such that $\Xi\psi = 0$ then also $X\xi = 0$, where ξ is the matrix $\binom{0}{\psi}$, which gives $X^2\xi = 0$; and it follows that the scalar square of the vector is zero. The determinant of Ξ is thus a power of $x^1x^{1'} + \cdots + x^\nu x^{\nu'}$ multiplied by a numerical factor which can only be ± 1 since in each row of Ξ, x^1 or $x^{1'}$ can occur, but not both. If the vector is a unit vector, the determinant of Ξ is thus always equal to $+1$ or always equal to -1. The matrix product of two unit vectors

$$AB = \begin{pmatrix} 0 & \alpha \\ \alpha^{-1} & 0 \end{pmatrix} \begin{pmatrix} 0 & \beta \\ \beta^{-1} & 0 \end{pmatrix} = \begin{pmatrix} \alpha\beta^{-1} & 0 \\ 0 & \alpha^{-1}\beta \end{pmatrix}$$

thus has the property that the two matrices of which it is composed are unimodular; the same result holds for the product matrix of any even number of unit matrices.

We finally note that the determinant of the matrix Ξ of a unit vector will be $+1$ or -1 depending on the order in which the compound indices of a spinor are arranged.

IV. SPECIAL CASES $\nu = 3$ AND $\nu = 4$

137. The case when $\nu = 3$

In the following chapter we shall examine in detail the case $\nu = 2$ which is of interest in Quantum Mechanics. The cases $\nu = 3$ and $\nu = 4$ call for some interesting geometric remarks.

For $\nu = 3$ the components $\xi_0, \xi_{23}, \xi_{31}, \xi_{12}$ of a semi-spinor of the first type can be regarded as the homogeneous co-ordinates of a point in a projective space of three dimensions, and the components (Section 134) $\xi_{123}, -\xi_1, -\xi_2, -\xi_3$ of a semi-spinor of the second type as the homogeneous co-ordinates of a plane in the same space. Equating the quantity

$$\varphi^T C\psi = \xi_0\xi_{123} - \xi_{23}\xi_1 - \xi_{31}\xi_2 - \xi_{12}\xi_3$$

to zero, gives the condition that the point is in the plane. The quantity $\varphi^T CX\varphi'$

defines an isotropic vector in the Euclidean space E_6; its covariant components are

$$x_1 = \xi_{31}\xi'_{12} - \xi_{12}\xi'_{31}, \quad x_2 = \xi_{12}\xi'_{23} - \xi_{23}\xi'_{12}, \quad x_3 = \xi_{23}\xi'_{31} - \xi_{31}\xi'_{23};$$
$$x'_1 = \xi_0\xi'_{23} - \xi_{23}\xi'_0, \quad x'_2 = \xi_0\xi'_{31} - \xi_{31}\xi'_0, \quad x'_3 = \xi_0\xi'_{12} - \xi_{12}\xi'_0.$$

These components are the Plücker co-ordinates of the line which joins the two points φ and φ' in three-dimensional space. In the same way the quantity $\psi^T C X \psi'$ defines an isotropic vector in E_6, with components

$$x_1 = \xi_{123}\xi'_1 - \xi_1\xi'_{123}, \quad x_2 = \xi_{123}\xi'_2 - \xi_2\xi'_{123}, \quad x_3 = \xi_{123}\xi'_3 - \xi_3\xi'_{123};$$
$$x'_1 = \xi_3\xi'_2 - \xi_2\xi'_3, \quad x'_2 = \xi_1\xi'_3 - \xi_3\xi'_1, \quad x'_3 = \xi_2\xi'_1 - \xi_1\xi'_2;$$

the components of this vector are the Plücker co-ordinates of the line of intersection of the two planes ψ and ψ' in three-dimensional space.

All semi-spinors are pure. A semi-spinor of the first type φ determines an isotropic triplane in E_6, that is, in the space of three dimensions, a two-parameter family of straight lines: this is the sheaf of lines passing through the image point of the semi-spinor φ. A semi-spinor of the second type ψ determines in the same way a two-parameter family of lines situated in the image plane of ψ. The isotropic triplanes associated with two semi-spinors of the same type must have in E_6 a line in common (Section 124); this corresponds to the obvious theorem that two points in three-dimensional space, or two planes in that space, determine a line. The theorem which states that the tri-plane associated with two semi-spinors of different types have either no line in common in E_6 or a biplane in common, corresponds to the theorem that a planes associated with two semi-spinors of different types have either no line straight line, but if the point lies in the plane they determine a pencil of lines.

We add that the group is of $(6 \times 5)/2 = 15$ parameters; the matrix Σ related to a rotation is the most general unimodular matrix of degree 4.

138. The case when $\nu = 4$; The fundamental trilinear form

In the case $\nu = 4$, semi-spinors of each type have eight components; each type has a quadratic form which is invariant with respect to the group of rotations, namely

$$\xi_0\xi_{1234} - \xi_{23}\xi_{14} - \xi_{31}\xi_{24} - \xi_{12}\xi_{34} \tag{4}$$

for semi-spinors of the first type, and

$$\xi_1\xi_{234} - \xi_2\xi_{134} + \xi_3\xi_{124} - \xi_4\xi_{123} \tag{5}$$

for semi-spinors of the second type.

We have here three spaces each of eight dimensions, that of vectors, that of semi-spinors of the first type, and that of semi-spinors of the second type, each having a fundamental quadratic form and in which there are three groups of operations which are the same overall, but with two to two correspondences which are not one-one, since to an operation in one of them there correspond two distinct operations in each of the others (Section 135).

These three groups, considered as acting simultaneously on vectors and on the two types of semi-spinors, form a group G which leaves invariant the trilinear form

$$\begin{aligned}\mathscr{F} = \varphi^{\mathrm{T}}\mathrm{C}\mathrm{X}\psi = x^1(&\xi_{12}\xi_{314} - \xi_{31}\xi_{124} - \xi_{14}\xi_{123} + \xi_{1234}\xi_1)\\ &+ x^2(\xi_{23}\xi_{124} - \xi_{12}\xi_{234} - \xi_{24}\xi_{123} + \xi_{1234}\xi_2)\\ &+ x^3(\xi_{31}\xi_{234} - \xi_{23}\xi_{314} - \xi_{34}\xi_{123} + \xi_{1234}\xi_3)\\ &+ x^4(-\xi_{14}\xi_{234} - \xi_{24}\xi_{314} - \xi_{34}\xi_{124} + \xi_{1234}\xi_4)\\ &+ x^{1'}(-\xi_0\xi_{234} + \xi_{23}\xi_4 - \xi_{24}\xi_3 + \xi_{34}\xi_2)\\ &+ x^{2'}(-\xi_0\xi_{314} + \xi_{31}\xi_4 - \xi_{34}\xi_1 + \xi_{14}\xi_3)\\ &+ x^{3'}(-\xi_0\xi_{124} + \xi_{12}\xi_4 - \xi_{14}\xi_2 + \xi_{24}\xi_1)\\ &+ x^{4'}(\xi_0\xi_{123} - \xi_{23}\xi_1 - \xi_{31}\xi_2 - \xi_{12}\xi_3),\end{aligned} \tag{6}$$

and the three quadratic forms

$$\left.\begin{aligned}\mathrm{F} &\equiv x^1x^{1'} + x^2x^{2'} + x^3x^{3'} + x^4x^{4'}\\ \Phi &\equiv \varphi^{\mathrm{T}}\mathrm{C}\varphi \equiv \xi_0\xi_{1234} - \xi_{23}\xi_{14} - \xi_{31}\xi_{24} - \xi_{12}\xi_{34}\\ \Psi &= \psi^{\mathrm{T}}\mathrm{C}\psi \equiv -\xi_1\xi_{234} - \xi_2\xi_{314} - \xi_3\xi_{124} + \xi_4\xi_{123}\end{aligned}\right\}. \tag{7}$$

Conversely any linear substitutions of the twenty-four variables x and ξ which transforms the vectors and semi-spinors of each type amongst themselves so as to leave the trilinear form $\mathscr{F}$ invariant will reproduce each of the forms F, Φ, Ψ to within a constant factor. Let us show this for the form F; the results stated below imply the proof for each of the two other forms Φ and Ψ. The invariance to within a factor of the form F follows from the characteristic property of an isotropic vector X of making the form $\mathscr{F}$, which is bilinear in the φ_α and ψ_β, degenerate. In fact saying that this form is degenerate is the same as saying that it is possible to find a semi-spinor ψ such that the form $\mathscr{F}$, linear in φ_α, is identically zero; the necessary and sufficient condition for this to hold is that $\mathrm{CX}\psi = 0$ or $\mathrm{X}\psi = 0$, and for this equality to hold for a semi-spinor ψ different from zero it is necessary that $\mathrm{X}^2\psi = 0$ or $\mathrm{X}^2 = 0$, and this is sufficient. The result is thus established.

We now take a linear substitution which leaves the forms $\mathscr{F}$, F, Φ, and Ψ invariant; then it is in the group G. To show this it is sufficient to show that, if it leaves all vectors invariant, it reduces to $\varphi' = \varphi$, $\psi' = \psi$ or to $\varphi' = -\varphi$, $\psi' = -\psi$. Now if the x^α are invariant, the coefficients of the x^α in the form $\mathscr{F}$ will also be invariant. Since $x^{1'}$ is invariant the transform of ξ_0 must depend only on $\xi_0, \xi_{23}, \xi_{24}, \xi_{34}$; invariance of $x^{2'}$ shows this transform depends only on $\xi_0, \xi_{31}, \xi_{34}, \xi_{14}$, and similarly for $x^{3'}$ and $x^{4'}$; it follows from this that the transform must be a multiple of ξ_0 and similarly for the other components; but it is easily seen that these multiples are the same for all components φ_2 and also the same (the inverse of the preceding constants) for the components ψ_β. The invariance of Φ and Ψ then shows that the constants are all $+1$ or all -1. In the same way it can be seen that the invariance of all the components φ_α implies either $x' = x$, $\psi' = \psi$ or $x' = -x$, $\psi' = -\psi$.

139. Principle of triality

We shall now show that the group G can be completed by five other families of linear substitutions which leave the form $\mathscr{F}$ invariant and interchange the three forms F, Φ, Ψ; the operations of each of these new families gives a definite permutation of the three sorts of object, vectors, semi-spinors of the first type, and semi-spinors of the second type.

One of these families, which we denote by $G_{(23)}$, is obtained by combining the operations of G with a reflection A which transforms X into $-AXA$, φ into $A\psi$, ψ into $-A\varphi$; thus for example take $A = H_{4'} + H_4$; this gives the operation

$$x^1 \to x^1, x^2 \to x^2, x^3 \to x^3, x^4 \to -x^{4'};$$

$$x^{1'} \to x^{1'}, x^{2'} \to x^{2'}, x^{3'} \to x^{3'}, x^{4'} \to -x^4;$$

$$\xi_0 \to \xi_4, \xi_{23} \to \xi_{234}, \xi_{31} \to \xi_{314}, \xi_{12} \to \xi_{124};$$

$$\xi_{14} \to \xi_1, \xi_{24} \to \xi_2, \xi_{34} \to \xi_3, \xi_{1234} \to \xi_{123}$$

$$\xi_1 \to -\xi_{14}, \xi_2 \to -\xi_{24}, \xi_3 \to \xi_{34}, \xi_4 \to -\xi_0;$$

$$\xi_{234} \to -\xi_{23}, \xi_{314} \to -\xi_{31}, \xi_{124} \to -\xi_{12}, \xi_{123} \to -\xi_{1234}.$$

This family is just that of reversals.

A second family which we shall call $G_{(12)}$ transforms vectors and semi-spinors of the first type into each other; it is obtained by combining G with the operation

$$x^1 \to -\xi_{23}, x^2 \to -\xi_{31}, x^3 \to -\xi_{12}, x^4 \to -\xi_0,$$

$$x^{1'} \to \xi_{14}, x^{2'} \to \xi_{24}, x^{3'} \to \xi_{34}, x^{4'} \to -\xi_{1234};$$

$$\xi_0 \to x^4, \xi_{23} \to x^1, \xi_{31} \to x^2, \xi_{12} \to x^3;$$

$$\xi_{14} \to -x^{1'}, \xi_{24} \to -x^{2'}, \xi_{34} \to -x^{3'}, \xi_{1234} \to x^{4'};$$

$$\xi_1 \to \xi_1, \xi_2 \to \xi_2, \xi_3 \to \xi_3, \xi_4 \to -\xi_{1234};$$

$$\xi_{234} \to \xi_{234}, \xi_{314} \to \xi_{314}, \xi_{124} \to \xi_{124}, \xi_{123} \to -\xi_4.$$

A third family, which we shall call $G_{(13)}$, transforms vectors and semi-spinors of the second type into each other; it is obtained by combining G with the operation

$$x^1 \to \xi_{234}, x^2 \to \xi_{314}, x^3 \to \xi_{124}, x^4 \to -\xi_{123};$$

$$x^{1'} \to -\xi_1, x^{2'} \to -\xi_2, x^{3'} \to -\xi_3, x^{4'} \to -\xi_4;$$

$$\xi_0 \to -\xi_{1234}, \xi_{23} \to \xi_{23}, \xi_{31} \to \xi_{31}, \xi_{12} \to \xi_{12};$$

$$\xi_{14} \to \xi_{14}, \xi_{24} \to \xi_{24}, \xi_{34} \to \xi_{34}, \xi_{1234} \to -\xi_0;$$

$$\xi_1 \to x^{1'}, \xi_2 \to x^{2'}, \xi_3 \to x^{3'}, \xi_4 \to x^{4'};$$

$$\xi_{234} \to -x^1, \xi_{314} \to -x^2, \xi_{124} \to -x^3, \xi_{123} \to x^4.$$

Finally the product $G_{(12)}G_{(13)}$ generates a fourth family, $G_{(132)}$, and the product $G_{(13)}G_{(12)}$ a fifth family, $G_{(123)}$; for example one of the operations of $G_{(123)}$ is

$$x^1 \to -\xi_{23}, x^2 \to -\xi_{31}, x^3 \to -\xi_{12}, x^4 \to \xi_{1234};$$
$$x^{1'} \to \xi_{14}, x^{2'} \to \xi_{24}, x^{3'} \to \xi_{34}, x^{4'} \to \xi_0;$$
$$\xi_0 \to -\xi_{123}, \xi_{23} \to \xi_{234}, \xi_{31} \to \xi_{314}, \xi_{12} \to \xi_{124};$$
$$\xi_{14} \to \xi_1, \xi_{24} \to \xi_2, \xi_{34} \to \xi_3, \xi_{1234} \to -\xi_4;$$
$$\xi_1 \to x^{1'}, \xi_2 \to x^{2'}, \xi_3 \to x^{3'}, \xi_4 \to -x^{4'};$$
$$\xi_{234} \to -x^1, \xi_{314} \to -x^2, \xi_{124} \to -x^3, \xi_{123} \to -x^{4'}.$$

We thus see that there is, in the geometry of eight-dimensional Euclidean space about a point, a *principle of triality** with three types of objects (vectors, semi-spinors of the first type, semi-spinors of the second type) which play exactly the same rôle. The group of this geometry is composed of six distinct continuous families corresponding to the six possible permutations of these three classes of objects; it is characterised by the invariance of $\mathscr{F}$ and of the form $F + \Phi + \Psi$.

140. Parallelism in the space E_8

To a unit bivector determined by two unit orthogonal vectors A, A′, we can associate, in the Euclidean space of semi-spinors of a given type, defined by the fundamental form Φ or Ψ, a paratactic congruence of unit bi-semi-spinors φ, φ' such that there exists one and only one of these bi-semi-spinors of which the first semi-spinor is a given φ. We shall define a unit semi-spinor as one which makes the fundamental form Φ equal to unity.

Note, now, that the matrix $X\varphi$, in which neither of the factors is zero, can only be zero if both the vector X and the semi-spinor φ are isotropic (i.e., of zero length). The first point has already been indicated; we can suppose that X has been reduced to the vector H_1; the equality $H_1\xi = 0$ implies the vanishing of all those components ξ_α which contain the index 1, and as a result the vanishing of the expression $\xi^T C\xi$, that is of the fundamental form Φ.

We now take φ to be a unit semi-spinor; let us find a semi-spinor φ' such that

$$(A + iA')(\varphi + i\varphi') = 0, \qquad (A - iA')(\varphi - i\varphi') = 0;$$

or

$$A\varphi - A'\varphi' = 0 \qquad A\varphi' + A'\varphi = 0;$$

the first equation gives $\varphi' = A'A\varphi$, and on substituting in the second we obtain an identity because $AA' + A'A = 0$. The remark made above now shows that

* Cf. E. Cartan, "Le principe de dualité et la théorie des groupes simples et semi-simples", *Bull. Sc. Math.*, **49**, 1925, 367–374.

$\varphi + i\varphi'$ and $\varphi - i\varphi'$ are isotropic semi-spinors and it follows that φ' is a unit semi-spinor orthogonal to φ.

If we use the language of projection geometry, the three spaces of vectors and of semi-spinors become seven-dimensional spaces, the line AA′ in the first space determines in each of the two others a paratactic congruence of straight lines such that one and only one passes through any point not on the fundamental quadric.

Given the unit vector A, this will allow us to define an "equipollence of unit bi-semi-spinors" in the space of semi-spinors φ (or ψ). Thus let (φ, φ') be a given unit bi-semi-spinor; the equalities

$$\mathrm{A}\varphi - \mathrm{A}'\varphi' = 0, \qquad \mathrm{A}\varphi' + \mathrm{A}'\varphi = 0,$$

where A, φ and φ' are given, have a definite solution for A′; this depends on the symmetric rôle of vectors and semi-spinors*. This being so, we shall say that the two unit bi-semi-spinors (φ, φ'), (φ_1, φ'_1) are equipollent if we find the same vector A′ for both. In the language of projective geometry, we see that, given a point in one of the three spaces (assumed not to lie on the fundamental quadric), it defines in the two other spaces a parallelism of straight lines (not on the fundamental quadric and not tangential to that quadric) such that through a given point passes one and only one parallel to a given straight line and two lines both parallel to a third are parallel to each other†.

141. Brioschi's formula

It is possible to deduce from the preceding considerations the formula due to Brioschi which gives the product of two sums each of eight squares in the form of a sum of eight squares. Let X by a vector, φ a semi-spinor; the product Xφ represents a semi-spinor ψ. We have

$$\psi^{\mathrm{T}}\mathrm{C}\psi = \varphi^{\mathrm{T}}\mathrm{X}^{\mathrm{T}}\mathrm{C}\mathrm{X}\varphi = \varphi^{\mathrm{T}}\mathrm{C}\mathrm{X}^2\varphi,$$

which gives the result that the scalar square of ψ is equal to the product of the scalar square of X and the scalar square of φ. Note that the components of ψ are bilinear forms in the components of X and the components of ψ. The formula can be expanded to give

$$\begin{aligned}
(x^1x^{1'} + x^2x^{2'} + x^3x^{3'} + x^4x^{4'})&(\xi_0\xi_{1234} - \xi_{23}\xi_{14} - \xi_{31}\xi_{24} - \xi_{12}\xi_{34}) \\
= (\xi_0x^{1'} - \xi_{31}x^3 + \xi_{12}x^2 + \xi_{14}x^4)&(\xi_{1234}x^1 - \xi_{23}x^{4'} + \xi_{24}x^{3'} - \xi_{34}x^{2'}) \\
+ (\xi_0x^{2'} - \xi_{12}x^1 + \xi_{23}x^3 + \xi_{24}x^4)&(\xi_{1234}x^2 - \xi_{31}x^{4'} + \xi_{34}x^{1'} - \xi_{14}x^{3'}) \\
+ (\xi_0x^{3'} - \xi_{23}x^2 + \xi_{31}x^1 + \xi_{34}x^4)&(\xi_{1234}x^3 - \xi_{12}x^{4'} + \xi_{14}x^{2'} - \xi_{24}x^{11}) \\
+ (\xi_0x^{4'} - \xi_{14}x^1 - \xi_{24}x^2 - \xi_{34}x^3)&(\xi_{1234}x^4 + \xi_{23}x^{1'} + \xi_{31}x^{2'} + \xi_{12}x^{3'}).
\end{aligned}$$

* We can also say that in the given vector space, the semi-spinors $\varphi + i\varphi'$ and $\varphi - i\varphi'$ are pure, and the two associated 4-vectors have no vector in common: we can then construct through the line in the direction of the vector A one and only one biplane which cuts each of the two 4-vectors along a line; these two lines are in the direction of A + iA′, A − iA′.

† See F. Vaney, "Le parallélisme absolu, etc.". Gauthier-Villars, Paris, 1929.

If we put $x^{1'}, x^{2'}, x^{3'}, x^{4'}$ as complex conjugates of x^1, x^2, x^3, x^4, and ξ_{1234}, $-\xi_{14}, -\xi_{24}, -\xi_{34}$ as complex conjugates of $\xi_0, \xi_{23}, \xi_{31}, \xi_{12}$, respectively, it can be seen that in each of the four products on the right-hand side the two factors are complex conjugates. We thus obtain Brioschi's formula *in the real domain*; the product of two sums of eight squares is itself the sum of eight squares. On putting

$$x^1 = X_0 + iX_1, \quad x^2 = X_4 + iX_3, \quad x^3 = X_6 + iX_2, \quad x^4 = X_7 + iX_5$$

$$\xi_0 = Y_0 + iY_2, \quad \xi_{14} = Y_7 - iY_5, \quad \xi_{24} = Y_6 + iY_2, \quad \xi_{34} = -Y_4 - iY_3$$

we obtain the formula

$$\begin{aligned}(X_0^2 + X_1^2 + X_2^2 + X_3^2 + X_4^2 + X_5^2 + X_6^2 + X_7^2)\\ \times (Y_0^2 + Y_1^2 + Y_2^2 + Y_3^2 + Y_4^2 + Y_5^2 + Y_6^2 + Y_7)^2\\ = Z_0^2 + Z_1^2 + Z_2^2 + Z_3^2 + Z_4^2 + Z_5^2 + Z_6^2 + Z_7^2,\end{aligned}$$

where we have written

$$\begin{aligned}Z_0 &= X_0Y_0 + X_1Y_1 + X_2Y_2 + X_3Y_3 + X_4Y_4 + X_5Y_5 + X_6Y_6 + X_7Y_7,\\ Z_i &= X_0Y_i - Y_0X_i + X_{i+1}Y_{i+5} - X_{i+5}Y_{i+1} + X_{i+2}Y_{i+3} - X_{i+3}Y_{i+2}\\ &\quad + X_{i+4}Y_{i+6} - X_{i+6}Y_{i+4}\end{aligned}$$

the indices $i + 1$, $i + 2$, $i + 3$, $i + 4$, $i + 5$, $i + 6$ are reduced (mod. 7) $(i = 1, 2, \ldots 7)$.

An application of Brioschi's formula to two unit vectors **X** and **Y**, gives a representation on a spherical space of 7 dimensions of the topological product of two spherical spaces of 7 dimensions.

We add one further remark. By setting $X_0 = Y_0 = 0$ we obtain a Euclidean space of 7 dimensions. The vector **Z** with components $z_1, z_2, \ldots, z_7$ is associated with the bivector formed from the two orthogonal unit vectors **X** and **Y**; it is also perpendicular to the bivector as can be shown by a simple calculation. The three vectors **X**, **Y**, and **Z** generate a triplane in which each vector is associated with the biplane to which it is perpendicular. Through any biplane there passes one and only one triplane of this type.

V. CASE OF REAL EUCLIDEAN SPACE

142. The matrices associated with real vectors

We can take the co-ordinates $x^i, x^{i'}$ to be complex conjugates. We already know (Section 112) that the matrix X associated with a real vector is Hermitian, which amounts to saying that the matrix H of degree $2^{\nu-1}$ equals the transpose of the conjugate of Ξ. If the vector is a unit vector, H is the inverse of Ξ, which is thus a unitary matrix. It follows easily from this, that in the matrix

$$\begin{pmatrix}\Sigma & 0\\ 0 & T\end{pmatrix}$$

representing a rotation the two component matrices are unitary and unimodular.

143. Conjugate spinors

As in the real space $E_{2\nu+1}$, we can take the spinor conjugate to ξ as $\xi' = i^\nu C\bar{\xi}$; ξ' transforms under a reflection A as a spinor if ν is even, but not if ν is odd (Section 114).

From this it follows that the quantities $\bar{\xi}^T \underset{(p)}{X} \xi$ provide a p-vector or an $(n-p)$-vector according to whether p is even or odd (Section 114). The scalar $\bar{\xi}^T\xi$ is just the sum of squares of the moduli of the ξ_α.

The tensor $\bar{\xi}_\alpha\xi_\beta$ decomposes into a scalar, a vector, a bivector, etc., an n-vector.

144. Conjugate semi-spinors

The conjugate of a semi-spinor is of the same type if ν is even, of a different type if ν is odd.

For odd ν the product of a semi-spinor by its conjugate decomposes into a scalar, a bivector, ..., a $(\nu-1)$-vector, all real, For even ν the decomposition also includes a scalar, a bivector, etc., but terminates with a semi-ν-vector.

145. Parallelisms in the elliptic space of 7 dimensions

For $\nu = 4$, the three spaces of vectors, of semi-spinors of the first type, and of semi-spinors of the second type, are real Euclidean spaces. Any real unit vector in one of these spaces provides in either of the two others an equipollence of unit bivectors. From the standpoint of projective geometry, this implies, in the elliptic (real projective) space of seven dimensions, the existence of a double infinity of parallelisms for directed straight lines.

VI. CASE OF PSEUDO-EUCLIDEAN SPACES

146. Conjugate spinors

As in the space $E_{2\nu+1}$ we assume that the co-ordinates x^i and $x^{i'}$ $(i = 1, 2, \ldots, \nu-h)$ are complex conjugates and that the co-ordinates $x^{\nu-h+1}, \ldots, x^\nu; x^{(\nu-h+1)'}, \ldots, x^{\nu'}$ are real. The results stated for $E_{2\nu+1}$ hold without modification in $E_{2\nu}$. Thus the conjugate of a spinor ξ can be defined by $\xi' = i^{\nu-h} I\bar{\xi}$ where

$$I = (H_1 - H_{1'})(H_2 - H_{2'})\ldots(H_{\nu-h} - H_{(\nu-h)'}).$$

The conjugate of a spinor behaves as a spinor under a real space-like reflection provided that ν is of the same parity as h (Section 117).

We can here demonstrate that decomposition of the product of a spinor by its conjugate gives a scalar $\bar{\xi}^{\mathrm{T}}\mathrm{J}\xi$, a vector, . . . , an n-vector.

147. Conjugate semi-spinors

Two conjugate semi-spinors will be of the same or different types according to whether h is of the same or different parity to ν.

If h is even, the product of a semi-spinor by its conjugate decomposes into a scalar, a bivector, etc.; the last irreducible tensor belonging to the decomposition is a $(\nu - 1)$-vector if ν is odd, a semi-ν-vector if ν is even.

If h is odd, the product of a semi-spinor by its conjugate decomposes into a vector, a trivector, etc.; the last tensor is a $(\nu - 1)$-vector if ν is even, a semi-ν-vector if ν is odd.

For $\nu = 3$ all semi-spinors are pure. If $h = 1$ or 3, the vector determined by a semi-spinor and its conjugate lies in the common straight line of the two associated isotropic triplanes which are complex conjugates of each other; the vector is thus real and isotropic. If $h = 3$ the two conjugate semi-spinors can be identical, in which case the vector is zero. If $h = 2$ the necessary and sufficient condition for the two associated triplanes to have a straight line in common—and they will then have a common real isotropic biplane—is that the scalar $\bar{\xi}^{\mathrm{T}}\mathrm{J}\xi$ be zero; this gives

$$\bar{\xi}_0\xi_{23} + \bar{\xi}_{12}\xi_{31} - \bar{\xi}_{31}\xi_{12} - \bar{\xi}_{23}\xi_0 = 0.$$

CHAPTER VII

SPINORS IN THE SPACE OF SPECIAL RELATIVITY (MINKOWSKI SPACE). DIRAC'S EQUATIONS

I. THE GROUP OF ROTATIONS IN EUCLIDEAN SPACE OF FOUR DIMENSIONS

148. The matrices associated with a *p*-vector

We shall first take as the fundamental form

$$F = x^1 x^{1'} + x^2 x^{2'}.$$

The general formulae now give without difficulty the matrices associated with a vector, a bivector, a trivector, and a four-vector.

We have for a vector

$$X = \begin{pmatrix} 0 & 0 & x^1 & x^2 \\ 0 & 0 & x^{2'} & -x^{1'} \\ x^{1'} & x^2 & 0 & 0 \\ x^{2'} & -x^1 & 0 & 0 \end{pmatrix}; \tag{1}$$

for a bivector, we have

$$\underset{(2)}{X} = \begin{pmatrix} \frac{1}{2}(x^{11'} + x^{22'}) & x^{12} & 0 & 0 \\ x^{2'1'} & -\frac{1}{2}(x^{11'} + x^{22'}) & 0 & 0 \\ 0 & 0 & -\frac{1}{2}(x^{11'} - x^{22'}) & x^{1'2} \\ 0 & 0 & x^{2'1} & \frac{1}{2}(x^{11'} - x^{22'}) \end{pmatrix}; \tag{2}$$

for a trivector, we have

$$\underset{(3)}{X} = \frac{1}{2}\begin{pmatrix} 0 & 0 & x^{122'} & x^{11'2} \\ 0 & 0 & -x^{11'2'} & x^{1'22'} \\ x^{1'22'} & -x^{11'2} & 0 & 0 \\ x^{11'2'} & x^{122'} & 0 & 0 \end{pmatrix}; \tag{3}$$

finally for a 4-vector, we have

$$\underset{(4)}{X} = \frac{1}{4}\begin{pmatrix} x^{11'22'} & 0 & 0 & 0 \\ 0 & x^{11'22'} & 0 & 0 \\ 0 & 0 & -x^{11'22'} & 0 \\ 0 & 0 & 0 & -x^{11'22'} \end{pmatrix}. \tag{4}$$

149. Case of orthogonal co-ordinates

If we take orthogonal co-ordinates by replacing $x^1, x^{1'}, x^2, x^{2'}$ by

$$x^1 + ix^2, \qquad x^1 - ix^2, \qquad x^3 + ix^4, \qquad x^3 - ix^4,$$

for a vector we obtain the matrix

$$X = \begin{pmatrix} 0 & 0 & x^1 + ix^2 & x^3 + ix^4 \\ 0 & 0 & x^3 - ix^4 & -x^1 + ix^2 \\ x^1 - ix^2 & x^3 + ix^4 & 0 & 0 \\ x^3 - ix^4 & -x^1 - ix^2 & 0 & 0 \end{pmatrix}. \tag{5}$$

The matrix $\underset{(2)}{X}$ of a bivector is of the form

$$\underset{(2)}{X} = \begin{pmatrix} \underset{(2)}{\Xi} & 0 \\ 0 & \underset{(2)}{H} \end{pmatrix}, \tag{6}$$

where

$$\underset{(2)}{\Xi} = \begin{pmatrix} -i(x^{12} + x^{34}) & -(x^{31} + x^{24}) + i(x^{23} + x^{14}) \\ x^{31} + x^{24} + i(x^{23} + x^{14}) & i(x^{12} + x^{34}) \end{pmatrix} \tag{7}$$

is the matrix of a semi-bivector of the first type with components

$$x^{23} + x^{14}, \qquad x^{31} + x^{24}, \qquad x^{12} + x^{34},$$

and where

$$H = \begin{pmatrix} i(x^{12} - x^{34}) & -(x^{31} - x^{24}) - i(x^{23} - x^{14}) \\ x^{31} - x^{24} - i(x^{23} - x^{14}) & -i(x^{12} - x^{34}) \end{pmatrix} \tag{8}$$

is the matrix of a semi-bivector of the second type with components

$$x^{23} - x^{14}, \qquad x^{31} - x^{24}, \qquad x^{12} - x^{34}.$$

If, on the contrary, we take co-ordinates appropriate to special relativity, with the fundamental form

$$(x^1)^2 + (x^2)^2 + (x^3)^2 - C^2(x^4)^2,$$

we have, for a vector, the matrix

$$X = \begin{pmatrix} 0 & 0 & x^1 + ix^2 & x^3 + cx^4 \\ 0 & 0 & x^3 - cx^4 & -x^1 + ix^2 \\ x^1 - ix^2 & x^3 + cx^4 & 0 & 0 \\ x^3 - cx^4 & -x^1 - ix^2 & 0 & 0 \end{pmatrix}; \tag{9}$$

then the matrix

$$\underset{(2)}{\Xi} = \begin{pmatrix} -i(x^{12} - icx^{34}) & -(x^{31} - icx^{24}) + i(x^{23} - icx^{14}) \\ x^{31} - icx^{24} + i(x^{23} - icx^{14}) & i(x^{12} - icx^{34}) \end{pmatrix} \tag{10}$$

represents a semi-bivector of the first type with components

$$x^{23} - icx^{14}, \qquad x^{31} - icx^{24}, \qquad x^{12} - icx^{34},$$

and the matrix

$$\underset{(2)}{H} = \begin{pmatrix} i(x^{12} + icx^{34}) & -(x^{31} + icx^{24}) - i(x^{23} + icx^{24}) \\ x^{31} + icx^{24} - i(x^{23} + icx^{14}) & -i(x^{12} + icx^{34}) \end{pmatrix} \tag{11}$$

represents a semi-bivector of the second type with components

$$x^{23} + icx^{14}, \qquad x^{31} + icx^{24}, \qquad x^{12} + icx^{34};$$

the covariant components of these semi-bivectors are respectively

$$x_{23} + \frac{i}{c}x_{14}, \qquad x_{31} + \frac{i}{c}x_{24}, \qquad x_{12} + \frac{i}{c}x_{34};$$

$$x_{23} - \frac{i}{c}x_{14}, \qquad x_{31} - \frac{i}{c}x_{24}, \qquad x_{12} - \frac{i}{c}x_{34}.$$

150. The group of rotations in complex space

Any unit vector A is of the form

$$A = \begin{pmatrix} 0 & \alpha \\ \alpha^{-1} & 0 \end{pmatrix};$$

the matrix α is of degree 2 and has determinant equal to -1. The reflection associated with this vector is given in terms of the semi-spinors φ and ψ by the relations

$$\varphi' = \alpha\psi, \qquad \psi' = \alpha^{-1}\varphi,$$

and in terms of the vector $X = \begin{pmatrix} 0 & \Xi \\ H & 0 \end{pmatrix}$ by the formulae

$$\Xi' = -\alpha H \alpha$$
$$H' = -\alpha^{-1}\Xi\alpha^{-1}.$$

The product of two reflections A and B gives the relations

$$\left.\begin{aligned} \varphi' &= \beta\alpha^{-1}\varphi, \qquad \psi' = \beta^{-1}\alpha\psi, \\ \Xi' &= \beta\alpha^{-1}\Xi\alpha^{-1}\beta \\ H' &= \beta^{-1}\alpha H\alpha\beta^{-1}. \end{aligned}\right\} \tag{12}$$

Put $\beta\alpha^{-1} = S$, $\beta^{-1}\alpha = t$; s and t are two unimodular matrices; this gives

$$\varphi' = s\varphi, \qquad \psi' = t\psi; \tag{13}$$

$$\Xi' = s\Xi t^{-1}, \qquad H' = tHs^{-1}; \tag{14}$$

these formulae apply for a simple rotation. Those for a general rotation are obviously of the same form. In particular the effect of any rotation on a vector X is given by a formula of the form

$$X' = \begin{pmatrix} s & 0 \\ 0 & t \end{pmatrix} X \begin{pmatrix} s^{-1} & 0 \\ 0 & t^{-1} \end{pmatrix} = SXS^{-1}, \tag{15}$$

where s and t are two unimodular matrices of degree 2.

Conversely, let s and t be any two unimodular matrices of degree 2; then the relations (15) define a rotation in the space of four dimensions. For this we must show that:

(i) If X is a matrix associated with a vector, then X′ is also a matrix associated with a vector.

(ii) The scalar square of the vector X′ is equal to the scalar square of the vector X.

The second proposition is obvious, since we have

$$X'^2 = SX^2S^{-1} = \mathbf{x}^2SS^{-1} = \mathbf{x}^2.$$

As to the first, it is true if X is the matrix of a unit vector, that is, if the determinant of Ξ equals -1 and if H is the inverse of Ξ, for then by (14) the determinant of Ξ' also equals -1, and H′ is obviously the inverse of Ξ'. The general case follows immediately from the special case.

THEOREM. *The most general rotation in four-dimensional complex Euclidean space is given by the formulae*

$$\xi' = S\xi, \qquad X' = SXS^{-1} \tag{16}$$

where the matrix S *is of the form* $\begin{pmatrix} s & 0 \\ 0 & t \end{pmatrix}$, *with two arbitrary unimodular complex matrices* s *and* t, *both of degree* 2.

The group of rotations is what is known as the *direct product* of two groups of linear unimodular substitutions in two variables.

If in particular we consider the transform of a bivector $\begin{pmatrix} \underset{(2)}{\Xi} & 0 \\ 0 & \underset{(2)}{\mathrm{H}} \end{pmatrix}$, we obtain

$$\underset{(2)}{\Xi'} = \underset{(2)}{s\Xi s^{-1}}, \qquad \underset{(2)}{\mathrm{H}'} = \underset{(2)}{t\mathrm{H}t^{-1}}. \tag{17}$$

Each of the two types of semi-bivector transforms according to a group isomorphic with the group of rotations in complex three-dimensional space. Note that the formulae (17) show that the determinant of each matrix $\underset{(2)}{\Xi}$ and $\underset{(2)}{\mathrm{H}}$ is an invariant of the group of rotations; these invariants are found to be the expressions

$$-\tfrac{1}{4}(x^{11} + x^{22'})^2 + x^{12}x^{1'2'}, \qquad -\tfrac{1}{4}(x^{11'} - x^{22'})^2 + x^{12'}x^{1'2}; \tag{18}$$

these are quadratic forms in the components of the corresponding semi-bivectors, which explains the result which we have just found. In orthogonal co-ordinates we have the two invariants

$$\left.\begin{aligned} &(x^{23} + x^{14})^2 + (x^{31} + x^{24})^2 + (x^{12} + x^{24})^2, \\ &(x^{23} - x^{14})^2 + (x^{31} - x^{24})^2 + (x^{12} - x^{24})^2, \end{aligned}\right\} \tag{19}$$

from which we deduce by subtraction the invariant $x^{23}x^{14} + x^{31}x^{24} + x^{12}x^{34}$; if this is zero then the bivector is simple. In special relativity the invariants are

$$\left.\begin{aligned} &(x^{23} - icx^{14})^2 + (x^{31} - icx^{24})^2 + (x^{12} - icx^{34})^2 \\ &(x^{23} + icx^{14})^2 + (x^{31} + icx^{24})^2 + (x^{12} + icx^{34})^2 \end{aligned}\right\}. \tag{20}$$

In particular it is seen that any simple bivector can be represented by two vectors of the same length in three-dimensional space; the effect produced on the bivector by a rotation in four-dimensional space is equivalent to the effects produced on each of the two representative vectors by two rotations each independent of the other.

151. Case of real Euclidean space

The formula (5) shows that the matrix α which constitutes one of the components of the matrix A associated with a real unit vector is a unitary matrix of determinant -1; the matrix A is in fact Hermitian, so that $\bar{\alpha} = (\alpha^{\mathrm{T}})^{-1}$. The unimodular matrices $s = -\beta\alpha^{-1}$ and $t = -\beta^{-1}\alpha$ are therefore unitary, which gives the

THEOREM. *Any rotation in real four-dimensional Euclidean space is expressed by the formulae* (15), *where the matrices s and t are any unitary unimodular matrices whatsoever.*

In particular, the components $x^{23} + x^{14}, x^{31} + x^{24}, x^{12} + x^{34}$ of a semi-bivector undergo a real orthogonal substitution of determinant 1.

152. Case of the space of special relativity

In the case of the space of special relativity the matrix α which forms part of the matrix A associated with a real unit vector is obviously the conjugate of α^{-1}:

$$\bar{\alpha} = \alpha^{-1}.$$

The matrices s and t are thus complex unimodular matrices and complex conjugates of each other.

THEOREM. *The most general proper rotation in the space of special relativity is given by the formulae*

$$\xi' = S\xi, \qquad X' = SXS^{-1}, \qquad S = \begin{pmatrix} s & 0 \\ 0 & \bar{s} \end{pmatrix},$$

where s is an arbitrary unimodular complex matrix of degree 2, and $\bar{s}$ is its conjugate.

We see from this that *the group of proper Lorentz rotations is isomorphic with the group of unimodular linear complex transformations of two variables, that is, isomorphic with the group of rotations of complex three-dimensional Euclidean space**.

The effect of proper reversals on spinors is given by the relations

$$\varphi' = s\psi, \qquad \psi' = \bar{s}\varphi, \qquad |s| = -1.$$

II. DECOMPOSITION OF THE PRODUCT OF TWO SPINORS

153. The product of two different spinors

By the general theorem of Section 131, the product of two spinors decomposes into a scalar, a vector, a bivector, a trivector, and a 4-vector. The bivector itself decomposes into two semi-bivectors. The quantities $\xi^{T}\underset{(p)}{C X}\xi'$ provide the tensors in this decomposition. The following results are found:

(i) The scalar is equal to

$$\xi_0\xi'_{12} - \xi_{12}\xi'_0 + \xi_2\xi'_1 - \xi_1\xi'_2.$$

(ii) The covariant components of the trivector are

$$x_{122'} = \xi_1\xi'_{12} - \xi_{12}\xi'_1, \qquad x_{11'2} = \xi_2\xi'_{12} - \xi_{12}\xi'_2,$$

$$x_{1'22'} = -\xi_2\xi'_0 + \xi_0\xi'_2, \qquad x_{11'2'} = -\xi_0\xi'_1 + \xi_1\xi'_0.$$

* A semi-vector of the first (second) type of Einstein and Mayer is a set of two semi-spinors of the first (second) type; it has no purely geometrical definition; its transformation law is given by the matrix $s[\bar{s}]$. See A. Einstein and W. Mayer, "Semi-Vektoren und Spinoren", *Sitzungsb. Akad. Berlin*, 1932, 522–550, and J. A. Schouten, "Zur generellen Feldtheorie. Semi-Vektoren und Spinraum", *Zeitschr. für Physik*, **84**, 1933, 92–111.

(iii) The covariant components of the semi-bivector of the first type are

$$x_{12} = -\xi_{12}\xi'_{12}, \quad x_{1'2'} = -\xi_0\xi'_0, \quad x_{11'} + x_{22'} = -(\xi_0\xi'_{12} + \xi_{12}\xi'_0),$$

and those of the semi-bivector of the second type are

$$x_{1'2} = \xi_2\xi'_2, \quad x_{12'} = \xi_1\xi'_1, \quad x_{11'} - x_{22'} = -(\xi_1\xi'_2 + \xi_2\xi'_1).$$

(iv) The covariant components of the vector are

$$x_1 = \tfrac{1}{2}(\xi_1\xi'_{12} + \xi_{12}\xi'_1), \qquad x_2 = \tfrac{1}{2}(\xi_2\xi'_{12} + \xi_{12}\xi'_2)$$

$$x'_1 = \tfrac{1}{2}(\xi_0\xi'_2 + \xi_2\xi'_0), \qquad x'_2 = -\tfrac{1}{2}(\xi_0\xi'_1 + \xi_1\xi'_0).$$

(v) Finally the 4-vector has its component

$$x_{11'22'} = \tfrac{1}{4}(\xi_0\xi'_{12} - \xi_{12}\xi'_0 + \xi_1\xi'_2 - \xi_2\xi'_1).$$

154. Semi-spinors defined as polarised isotropic bivectors

A less abstract definition of semi-spinors as polarised isotropic semi-bivectors is obtained by putting $\xi' = \xi$; this gives for an isotropic bivector of the first type

$$x_{12} = -\xi_{12}^2, \qquad x_{1'2'} = -\xi_0^2, \qquad x_{11'} + x_{22'} = -2\xi_0\xi_{12};$$

$$x_{1'2} = 0, \qquad x_{12'} = 0, \qquad x_{11'} - x_{22'} = 0,$$

and for an isotropic bivector of the second type*

$$x_{1'2} = \xi_2^2, \qquad x_{12'} = \xi_1^2, \qquad x_{11'} - x_{22'} = -2\xi_1\xi_2;$$

$$x_{12} = 0, \qquad x_{1'2'} = 0, \qquad x_{11'} + x_{22'} = 0.$$

In the space of special relativity we can give another interpretation of semi-spinors. We use the co-ordinates previously introduced: x^1, x^2, x^3, x^4. The semi-bivector of a spinor is given by the formulae

$$\left.\begin{aligned} cx^{14} + ix^{23} &= \tfrac{1}{2}(-\xi_0^2 + \xi_{12}^2),\\ cx^{24} + ix^{31} &= \frac{i}{2}(\xi_0^2 + \xi_{12}^2),\\ cx^{34} + ix^{12} &= -\xi_0\xi_{12}, \end{aligned}\right\} \qquad (21)$$

and

$$\left.\begin{aligned} cx^{14} - ix^{23} &= \tfrac{1}{2}(\xi_1^2 - \xi_2^2),\\ cx^{24} - ix^{31} &= \frac{i}{2}(\xi_1^2 + \xi_2^2),\\ cx^{34} - ix^{12} &= \xi_1\xi_2. \end{aligned}\right\} \qquad (22)$$

* E. T. Whittaker noticed this relation between certain bivectors in the space of special relativity and semi-spinors: "On the relations of the tensor-calculus to the spinor-calculus", *Proc. R. Soc. London*, **158**, 1937, 38–46.

The formulae (21) allow us to represent the semi-spinor (ξ_0, ξ_{12}) in terms of a *real* bivector x^{ij} which satisfies the two relations

$$\left.\begin{aligned} &c^2[(x^{14})^2 + (x^{24})^2 + (x^{34})^2] = (x^{23})^2 + (x^{31})^2 + (x^{12})^2, \\ &x^{14}x^{23} + x^{24}x^{31} + x^{34}x^{12} = 0; \end{aligned}\right\} \tag{23}$$

the semi-spinor is, then, the polarised bivector. Physically this bivector can be regarded as the combination of an electric field (x^{14}, x^{24}, x^{34}) and a magnetic field (x^{23}, x^{31}, x^{12}), at right angles to each other and with intensities in the ratio $1/c$, the reciprocal of the speed of light; this is the form in which the two fields occur in the electromagnetic theory of light. The semi-spinor (ξ_1, ξ_2) can be given an analogous interpretation using the formulae (22).

An interpretation of this sort in terms of a real image is possible only in the space of special relativity, but not in real Euclidean four-dimensional space.

III. CONJUGATE VECTORS AND SPINORS IN THE SPACE OF SPECIAL RELATIVITY

155. Conjugate spinors

The matrices associated with two conjugate imaginary vectors are

$$\begin{pmatrix} 0 & 0 & x^1 & x^2 \\ 0 & 0 & x^{2'} & -x^{1'} \\ x^{1'} & x^2 & 0 & 0 \\ x^{2'} & -x^1 & 0 & 0 \end{pmatrix} \quad \text{and} \quad \begin{pmatrix} 0 & 0 & \bar{x}^{1'} & \bar{x}^2 \\ 0 & 0 & \bar{x}^{2'} & -\bar{x}^1 \\ \bar{x}^1 & \bar{x}^2 & 0 & 0 \\ \bar{x}^{2'} & -\bar{x}^{1'} & 0 & 0 \end{pmatrix}$$

that is

$$\begin{pmatrix} 0 & \Xi \\ \mathrm{H} & 0 \end{pmatrix} \quad \text{and} \quad \begin{pmatrix} 0 & \bar{\mathrm{H}} \\ \bar{\Xi} & 0 \end{pmatrix}.$$

The latter Y can be deduced from the former X by the formula (Section 116)

$$\mathrm{Y} = -\mathrm{I}\bar{\mathrm{X}}\mathrm{I}^{\mathrm{T}} = (\mathrm{H}_1 - \mathrm{H}_{1'})\bar{\mathrm{X}}(\mathrm{H}_1 - \mathrm{H}_{1'}).$$

In the same way, for the spinor ξ' conjugate to a given spinor ξ we have the formula

$$\xi' = i(\mathrm{H}_1 - \mathrm{H}'_1)\bar{\xi} = i\begin{pmatrix} 0 & 1 \\ -1 & 0 \end{pmatrix}\bar{\xi},$$

which gives

$$\varphi' = i\bar{\Psi}, \qquad \psi' = -i\bar{\varphi},$$

or, in more detail

$$\xi'_0 = i\bar{\xi}_1, \qquad \xi'_{12} = i\bar{\xi}_2, \qquad \xi'_1 = -i\bar{\xi}_0, \qquad \xi'_2 = -i\bar{\xi}_{12}. \tag{24}$$

156. Decomposition of the product of two conjugate spinors

The results of Section 146 give immediately the decomposition of the product of two conjugate spinors into a scalar, a vector, a bivector, a trivector, and a four-vector.

We shall calculate these various tensors using the co-ordinates of special relativity, for which the fundamental form is

$$(x^1)^2 + (x^2)^2 + (x^3)^2 - c^2(x^4)^2;$$

the matrices associated with the basis vectors are respectively

$$\mathrm{A}_1 = \mathrm{H}_1 + \mathrm{H}_{1'}, \quad \mathrm{A}_2 = i(\mathrm{H}_1 - \mathrm{H}_{1'}), \quad \mathrm{A}_3 = \mathrm{H}_2 + \mathrm{H}_{2'}, \quad \mathrm{A}_4 = c(\mathrm{H}_2 - \mathrm{H}_{2'}).$$

Here the matrix J of Section 116 is $\mathrm{H}_2 - \mathrm{H}_{2'}$ and we shall consider the quantities $\bar\xi^{\mathrm{T}}\mathrm{J}\underset{(p)}{\mathrm{X}}\xi$, but we shall multiply each one by a constant factor so as to obtain real p-vectors.

(i) There is a four-vector with component

$$i\bar\xi^{\mathrm{T}}\mathrm{J}\xi = -i(\xi_0\bar\xi_2 - \xi_{12}\bar\xi_1 + \xi_1\bar\xi_{12} - \xi_2\bar\xi_0). \tag{25}$$

(ii) There is a vector given by the quantity $\bar\xi^{\mathrm{T}}\mathrm{J}\mathrm{X}\xi$, having as contravariant components

$$\left.\begin{aligned} x^1 &= -(\xi_0\bar\xi_{12} + \bar\xi_{12}\xi_0 + \xi_1\bar\xi_2 + \xi_2\bar\xi_1), \\ x^2 &= i(\xi_0\bar\xi_{12} - \bar\xi_{12}\xi_0 - \xi_1\bar\xi_2 + \xi_2\bar\xi_1), \\ x^3 &= \xi_0\bar\xi_0 - \xi_{12}\bar\xi_{12} + \xi_1\bar\xi_1 - \xi_2\bar\xi_2, \\ x^4 &= \frac{1}{c}(\xi_0\bar\xi_0 + \xi_{12}\bar\xi_{12} + \xi_1\bar\xi_1 + \xi_2\bar\xi_2). \end{aligned}\right\} \tag{26}$$

By a general theorem of Section 119, this is a time-like vector, of which the scalar square is

$$-4(\xi_0\bar\xi_2 - \xi_{12}\bar\xi_1)(\xi_2\bar\xi_0 - \xi_1\bar\xi_{12}).$$

(iii) There is a bivector given by the quantity $\bar\xi^{\mathrm{T}}\mathrm{J}\underset{(2)}{\mathrm{X}}\xi$, having as contravariant components

$$\left.\begin{aligned} x^{23} &= i(\xi_0\bar\xi_1 - \xi_{12}\bar\xi_2 - \xi_1\bar\xi_0 + \xi_2\bar\xi_{12}), \\ x^{31} &= \xi_0\bar\xi_1 + \xi_{12}\bar\xi_2 + \xi_1\bar\xi_0 + \xi_2\bar\xi_{12}, \\ x^{12} &= i(\xi_0\bar\xi_2 + \xi_{12}\bar\xi_1 - \xi_1\bar\xi_{12} - \xi_2\bar\xi_0), \\ x^{14} &= \frac{1}{c}(-\xi_0\bar\xi_1 + \xi_{12}\bar\xi_2 - \xi_1\bar\xi_0 + \xi_2\bar\xi_{12}), \\ x^{24} &= \frac{i}{c}(\xi_0\bar\xi_1 + \xi_{12}\bar\xi_2 - \xi_1\bar\xi_0 - \xi_2\bar\xi_{12}), \\ x^{34} &= -\frac{1}{c}(\xi_0\bar\xi_2 + \xi_{12}\bar\xi_1 + \xi_1\bar\xi_{12} + \xi_2\bar\xi_0). \end{aligned}\right\} \tag{27}$$

This decomposes into two conjugate imaginary semi-bivectors

$$\left.\begin{aligned} x^{23} - icx^{14} &= 2i(\xi_0\bar{\xi}_1 - \xi_{12}\bar{\xi}_2), \\ x^{31} - icx^{24} &= 2(\xi_0\bar{\xi}_1 + \xi_{12}\bar{\xi}_2), \\ x^{12} - icx^{34} &= 2i(\xi_0\bar{\xi}_2 + \xi_{12}\bar{\xi}_1); \\ x^{23} + icx^{14} &= 2i(-\xi_1\bar{\xi}_0 + \xi_2\bar{\xi}_{12}), \\ x^{31} + icx^{24} &= 2(\xi_1\bar{\xi}_0 + \xi_2\bar{\xi}_{12}), \\ x^{12} + icx^{34} &= -2i(\xi_1\bar{\xi}_{12} + \xi_2\bar{\xi}_0). \end{aligned}\right\} \tag{28}$$

The sum of the squares of the components of the first is equal to $-4(\xi_0\bar{\xi}_2 - \xi_{12}\bar{\xi}_1)^2$; the sum of the squares of the components of the second is $-4(\xi_2\bar{\xi}_0 - \xi_1\bar{\xi}_{12})^2$.

(iv) There is a trivector given by the quantity $i\bar{\xi}^{\mathrm{T}}\underset{(3)}{\mathrm{JX}}\xi$, with components

$$\left.\begin{aligned} x^{234} &= \frac{1}{c}(-\xi_0\bar{\xi}_{12} - \xi_{12}\bar{\xi}_0 + \xi_1\bar{\xi}_2 + \xi_2\bar{\xi}_1), \\ x^{314} &= \frac{i}{c}(\xi_0\bar{\xi}_{12} - \xi_{12}\bar{\xi}_0 + \xi_1\bar{\xi}_2 - \xi_2\bar{\xi}_1), \\ x^{124} &= \frac{1}{c}(\xi_0\bar{\xi}_0 - \xi_{12}\bar{\xi}_{12} - \xi_1\bar{\xi}_1 + \xi_2\bar{\xi}_2), \\ x^{123} &= \xi_0\bar{\xi}_0 + \xi_{12}\bar{\xi}_{12} - \xi_1\bar{\xi}_1 - \xi_2\bar{\xi}_2. \end{aligned}\right\} \tag{29}$$

(v) Finally there is a scalar

$$\xi_0\bar{\xi}_2 - \xi_{12}\bar{\xi}_1 - \xi_1\bar{\xi}_{12} + \xi_2\bar{\xi}_0. \tag{30}$$

IV. DIRAC'S EQUATIONS

157. The covariant vector $\partial/\partial x$

Consider in the space of special relativity referred to co-ordinates x^1, x^2, x^3, x^4, a function of position f; the differential $\mathrm{d}f$ is an invariant scalar under all direct or inverse Lorentz transformations; now

$$\mathrm{d}f \equiv \frac{\partial f}{\partial x^1}\mathrm{d}x^1 + \frac{\partial f}{\partial x^2}\mathrm{d}x^2 + \frac{\partial f}{\partial x^3}\mathrm{d}x^3 + \frac{\partial f}{\partial x^4}\mathrm{d}x^4;$$

the differentials $\mathrm{d}x^i$ transform as the components of a contravariant vector and consequently we can regard the four operators $\partial/\partial x^i$ as the components of a covariant vector; the contravariant components of this vector are

$$\frac{\partial}{\partial x^1}, \frac{\partial}{\partial x^2}, \frac{\partial}{\partial x^3}, -\frac{1}{c^2}\frac{\partial}{\partial x^4}.$$

We shall denote the associated matrix by $\partial/\partial x$, that is

$$\frac{\partial}{\partial x} = \begin{pmatrix} 0 & 0 & \frac{\partial}{\partial x^1} + i\frac{\partial}{\partial x^2} & \frac{\partial}{\partial x^3} - \frac{1}{c}\frac{\partial}{\partial x^4} \\ 0 & 0 & \frac{\partial}{\partial x^3} + \frac{1}{c}\frac{\partial}{\partial x^4} & -\frac{\partial}{\partial x^1} + i\frac{\partial}{\partial x^3} \\ \frac{\partial}{\partial x^1} - i\frac{\partial}{\partial x^2} & \frac{\partial}{\partial x^3} - \frac{1}{c}\frac{\partial}{\partial x^4} & 0 & 0 \\ \frac{\partial}{\partial x^3} + \frac{1}{c}\frac{\partial}{\partial x^4} & -\frac{\partial}{\partial x^1} - i\frac{\partial}{\partial x^2} & 0 & 0 \end{pmatrix}.$$

With this notation the Dirac equations for an electron in an electromagnetic field are as follows. We introduce four wave functions which are the components of a spinor ξ and functions of position (x); let V be the matrix associated with the vector potential and K the matrix already considered (Section 126)

$$\mathrm{K} = \begin{pmatrix} -1 & 0 & 0 & 0 \\ 0 & -1 & 0 & 0 \\ 0 & 0 & 1 & 0 \\ 0 & 0 & 0 & 1 \end{pmatrix}.$$

Dirac's equations are summarised in the equation

$$\left(\frac{h}{i}\frac{\partial}{\partial x} + \frac{e}{c}\mathrm{V} - m_0 c\mathrm{K}\right)\xi = 0,$$

where the symbols h, e, c, m_0 have well-known physical meanings. On writing out each of the four equations we obtain*

$$\begin{aligned}
&\frac{h}{i}\left(\frac{\partial \xi_1}{\partial x^1} + i\frac{\partial \xi_1}{\partial x^2} + \frac{\partial \xi_2}{\partial x^3} - \frac{1}{c}\frac{\partial \xi_2}{\partial x^4}\right) + \frac{e}{c}[(\mathrm{V}^1 + i\mathrm{V}^2)\xi_1 + (\mathrm{V}^3 + i\mathrm{V}^4)\xi_2] = -m_0 c\xi_0,\\
&\frac{h}{i}\left(\frac{\partial \xi_1}{\partial x^3} + \frac{1}{c}\frac{\partial \xi_1}{\partial x^4} - \frac{\partial \xi_2}{\partial x^1} + i\frac{\partial \xi_2}{\partial x^2}\right) + \frac{e}{c}[(\mathrm{V}^3 - c\mathrm{V}^4)\xi_1 - (\mathrm{V}^1 - i\mathrm{V}^2)\xi_2] = -m_0 c\xi_{12},\\
&\frac{h}{i}\left(\frac{\partial \xi_0}{\partial x^1} - i\frac{\partial \xi_0}{\partial x^2} + \frac{\partial \xi_{12}}{\partial x^3} - \frac{1}{c}\frac{\partial \xi_{12}}{\partial x^4}\right) + \frac{e}{c}[(\mathrm{V}^1 - i\mathrm{V}^2)\xi_0 + (\mathrm{V}^3 + c\mathrm{V}^4)\xi_{12}] = m_0 c\xi_1,\\
&\frac{h}{i}\left(\frac{\partial \xi_0}{\partial x^3} + \frac{1}{c}\frac{\partial \xi_0}{\partial x^4} - \frac{\partial \xi_{12}}{\partial x^1} - i\frac{\partial \xi_{12}}{\partial x^2}\right) + \frac{e}{c}[(\mathrm{V}^3 - c\mathrm{V}^4)\xi_0 - (\mathrm{V}^4 + i\mathrm{V}^2)\xi_{12}] = m_0 c\xi_2.
\end{aligned} \tag{31}$$

* See for example B. L. van der Waerden, "Die gruppentheoretische Methode in der Quantenmechanik", 1932, p. 97; we can pass from the equations (23.7) of that book to the formulae (31) of the text by replacing Ψ by $(\xi_1, -\xi_2)$ and Ψ by (ξ_{12}, ξ_0). We can pass from the notation in the text to that of Dirac by replacing ξ_0 by $\psi_2 - \psi_4$, ξ_1 by $\psi_1 + \psi_3$, ξ_2 by $-\psi_2 - \psi_4$, and ξ_{12} by $\psi_1 - \psi_3$.

158. The divergence of the current vector

It is now easy to show that, if the spinor ξ satisfies Dirac's equations, then the divergence of the vector (26) is zero. In Quantum Mechanics this vector plays the rôle of current vector. In fact the scalar product of an arbitrary real vector X by the current vector is the real quantity $\bar{\xi}^{T}JX\xi$. The divergence of the current vector is the sum of two conjugate imaginary quantities, one comes from the derivatives of the components ξ_{α}, the other from the derivatives of the components $\bar{\xi}_{\alpha}$. The first quantity is $\bar{\xi}^{T}J(\partial/\partial x)\xi$ which by Dirac's equations is equal to

$$-\frac{ei}{ch}\bar{\xi}^{T}JV\xi + \frac{m_0 ci}{h}\bar{\xi}^{T}JK\xi;$$

now the quantity $\bar{\xi}^{T}JV\xi$ is real since the vector V is real; in the same way the quantity $\bar{\xi}^{T}JK\xi$ is equal to the scalar (30)

$$\xi_0\bar{\xi}_2 + \xi_2\bar{\xi}_0 - \xi_1\bar{\xi}_{12} - \xi_{12}\bar{\xi}_1;$$

we thus obtain for $\bar{\xi}^{T}J(\partial/\partial x)\xi$ a purely imaginary value; the divergence we are looking for, the sum of two purely imaginary quantities, conjugates of each other, is thus zero.

Dirac's equations are invariant under all proper (direct or inverse) transformations of the Lorentz group, for under a real space reflection A, the two matrices $V\xi$ and $K\xi$ with one column and four rows transform in the same way; the first becomes $-A(V\xi)$, the second $KA\xi = -A(K\xi)$.

159. Note

In a space with an odd number of dimensions, systems of equations analogous to Dirac's equations and which are invariant under displacements and reversals do not exist; this follows from the fact that the spinor $(\partial/\partial x)\xi$ (or $V\xi$) is not equivalent to the spinor ξ with respect to reversals. But in a space with any even number of dimensions whatsoever, Dirac's equations generalise as they stand.

CHAPTER VIII

LINEAR REPRESENTATIONS OF THE LORENTZ GROUP

I. LINEAR REPRESENTATIONS OF THE GROUP OF LORENTZ ROTATIONS

160. Reduction to the group of complex rotations in E_3

The theorem of Section 75 shows that there exists a one to one correspondence between the irreducible linear representations of:

(i) The group of rotations in real Euclidean space of four dimensions.
(ii) The group of Lorentz rotations (proper rotations).
(iii) The group of proper rotations in the pseudo-Euclidean space which has fundamental form reducible to the sum of two positive and two negative squares.

By passage from real to complex, each of these three groups gives the same group, namely that of complex rotations.

In the first group, the semi-spinor (ξ_1, ξ_2) undergoes a unimodular unitary substitution and the semi-spinor (ξ_0, ξ_{12}) undergoes a substitution of the same type as, but independent of, the first.

In the second group, the semi-spinor (ξ_1, ξ_2) undergoes a unimodular linear substitution s with complex coefficients and the semi-spinor (ξ_0, ξ_{12}) undergoes the complex conjugate substitution $\bar{s}$ (Section 152).

In the third group, the two semi-spinors undergo independent real unimodular substitutions.

We have already determined the linear representations of the group of complex rotations in three-dimensional space (Sections 82–84) and this group is isomorphic to the Lorentz group. A complete system of non-equivalent irreducible representations is given by the representations $\mathscr{D}_{p/2,q/2}$ with generating polynomials

$$(a\xi_1 + b\xi_2)^p(c\xi_0 + d\xi_{12})^q.$$

161. Particular cases

For $p = q = 1$ we obtain a tensor of degree 4 which is equivalent to a vector; in fact the expression $\xi^T CX\xi'$, where ξ is a semi-spinor (ξ_1, ξ_2) and ξ' a semi-spinor (ξ_0, ξ_{12}), gives a vector with components

$$x_1 = x^{1'} = \xi_1\xi_{12}, \quad x_2 = x^{2'} = \xi_2\xi_{12}, \quad x_{1'} = x^1 = \xi_2\xi_0, \quad x_{2'} = x^2 = -\xi_1\xi_0.$$

The generating polynomial of this tensor can be written

$$F_{1,1} \equiv (a\xi_1 + b\xi_2)(c\xi_0 + d\xi_{12}) \sim bcx^1 - acx^2 + adx^{1'} + bdx^{2'}.$$

If we assume $p > q$, we can substitute, for the generating polynomial of $\mathscr{D}_{p/2,q/2}$, the form

$$F_{p,q} \sim (bcx^1 - acx^2 + adx^{1'} + bdx^{2'})^q(a\xi_1 + b\xi_2)^{p-q}.$$

For $p = q$ we obtain a tensor of degree $(p + 1)^2$; the components are homogeneous integral polynomials of degree p in $x^1, x^2, x^{1'}, x^{2'}$ which satisfy the Laplace equation

$$\frac{\partial^2 V}{\partial x^1 \partial x^{2'}} + \frac{\partial^2 V}{\partial x^2 \partial x^{2'}} = 0;$$

these are the harmonic polynomials of degree p; the proof is immediate.

For $p = 2$, $q = 0$, the tensor has as components $\xi_1^2, \xi_1\xi_2, \xi_2^2$; it is equivalent to a semi-bivector of the second type, and one can take as generating polynomial

$$F_{20} \sim a^2 x_{12'} - ab(x_{11'} - x_{22'}) + b^2 x_{1'2};$$

in the same way

$$F_{02} \sim c^2 x_{1'2'} + cd(x_{11'} + x_{22'}) + d^2 x_{12}.$$

If p and q have the same parity, we can take the generating form of $\mathscr{D}_{p/2,q/2}$ in terms of vectors and semi-bivectors only.

II. REPRESENTATIONS OF THE LORENTZ GROUP OF ROTATIONS AND REVERSALS

162. The two categories of irreducible representations

The group of rotations and reversals of real four-dimensional Euclidean space, the group of proper rotations and reversals in the space of special relativity, and finally the group of proper rotations and reversals in the real pseudo-Euclidean space with fundamental form reducible to the sum of two positive squares and two negative squares, all have the same linear representations.

In any one of them, the reflection $H_2 + H_{2'}$ transforms ξ_0 into ξ_2 and ξ_2 into ξ_0, and thus transforms the component $\xi_0^p\xi_2^q$ of the irreducible tensor $\mathscr{D}_{p/2,\,q/2}$ into the component $\xi_0^q\xi_2^p$ of the tensor $\mathscr{D}_{q/2,\,p/2}$. Application of the theorems of Sections 88 and 89 thus leads to the following theorem.

THEOREM. *The irreducible representations of the group of (proper) rotations and reversals in Euclidean space of four dimensions can be classified into two categories:*

(i) *Those which induce in the group of rotations one irreducible linear representation; this is necessarily of the form $\mathscr{D}_{p/2,\,p/2}$; to each value of p there correspond two non-equivalent irreducible linear representations of the full group.*

(ii) *Those which induce a reducible linear representation in the group of rotations; this decomposes into two non-equivalent irreducible representations $\mathscr{D}_{p/2,\,q/2}$ and $\mathscr{D}_{q/2,\,p/2}$ $(p \neq q)$. To each pair (p, q) of different integers corresponds a single irreducible representation of the full group.*

We shall denote by $\mathscr{D}^+_{(p/2,\,q/2)}$ the tensor with components which transform as do those of the tensor with generating form

$$(a\xi_1 + b\xi_2)^p(c\xi_0 + d\xi_{12})^p,$$

and by $\mathscr{D}^-_{(p/2,\,p/2)}$ the tensor which has components which transform in the same way except that they change sign under any reversal. For $p = 1$, $\mathscr{D}^+_{(\frac{1}{2},\,\frac{1}{2})}$ represents a vector, and thus it follows that $\mathscr{D}^-_{(\frac{1}{2},\,\frac{1}{2})}$ represents a trivector.

We denote by $\mathscr{D}_{(p/2,\,q/2)}$ $(p \neq q)$ the tensor with generating form

$$(a\xi_1 + b\xi_2)^p(c\xi_0 + d\xi_{12})^q + (a\xi_1 + b\xi_2)^q(c\xi_0 + d\xi_{12})^p.$$

163. Decomposition of the product $\mathscr{D}^+_{(p/2,\,p/2)} \times \mathscr{D}^+_{(p'/2,\,p'/2)}$

The components of the two tensors are provided by the generating polynomials

$$(a\xi_1 + b\xi_2)^p(c\xi_0 + d\xi_{12})^p,$$
$$(a\xi'_1 + b\xi'_2)^{p'}(c\xi'_0 + d\xi'_{12})^{p'}.$$

Form the polynomials

$$P_{i,j} = (\xi_1\xi'_2 - \xi_2\xi'_1)^i(\xi_0\xi'_{12} - \xi_{12}\xi'_0)^j(a\xi_1 + b\xi_2)^{p-i}(a\xi'_1 + b\xi'_2)^{p'-i}$$
$$\times (c\xi_0 + d\xi_{12})^{p-j}(c\xi'_0 + d\xi'_{12})^{p'-j} \quad (0 \leqslant i \leqslant p', 0 \leqslant j \leqslant p', p \geqslant p').$$

The coefficients of the various monomials in a, b, c, d give $(p + p' - 2i + 1)$ $(p + p' - 2j + 1)$ linear combinations of components of the product tensor. The sum of all such numbers equals the square of the sum of the quantities $p + p' - 2i + 1$ $(i = 0, 1, \ldots p')$; it is therefore equal to $(p + 1)^2(p' + 1)^2$. But this latter number is exactly the degree of the product tensor of which the decomposition is required. Each of the polynomials p_{ij} determines an irreducible tensor of the group of rotations, namely the tensor $\mathscr{D}_{(p+p'/2)-i,\,(p+p'/2)-j}$; we thus have the required decomposition

$$\mathscr{D}^+_{(p/2,\,p/2)} \times \mathscr{D}^+_{(p'/2,\,p'/2)} = \sum \mathscr{D}_{[(p+p'/2)-i,\,(p+p'/2)-j]} + \sum \mathscr{D}^+_{[(p+p'/2)-i,(p+p'/2)-i]};$$

the first sum extends over all pairs of integers i, j such that $0 \leqslant i < j \leqslant p' \leqslant p$, the second over all integers i such that $0 \leqslant i \leqslant p' \leqslant p$. The tensors in the second sum are $\mathscr{D}^+$, thus we have

$$\mathrm{P}_{i,i} = (\xi_1\xi'_2 - \xi_2\xi'_1)^i(\xi_0\xi'_{12} - \xi_{12}\xi'_0)^i(a\xi_1 + b\xi_2)^{p-i}(a\xi'_1 + b\xi'_2)^{p-i} \times (c\xi_0 + d\xi_{12})^{p'-i}(c\xi'_0 + d\xi'_{12})^{p'-i};$$

under a reversal the two quantities $(\xi_1\xi'_2 - \xi_2\xi'_1)$ and $(\xi_0\xi'_{12} - \xi_{12}\xi'_0)$ transform, with a change in sign, into each other; therefore the polynomial $\mathrm{P}_{i,c}$ is equivalent to the generating polynomial of $\mathscr{D}^+_{[(p+p'/2)-i,(p+p'/2)-i]}$.

If the two factors on the left-hand side are both $\mathscr{D}^-$, the result will be the same. If only one of these factors is a $\mathscr{D}^-$, then all the tensors in the second sum will be $\mathscr{D}^-$.

For example we have

$$\mathscr{D}^-_{(\frac{1}{2},\frac{1}{2})} \times \mathscr{D}^-_{(\frac{1}{2},\frac{1}{2})} = \mathscr{D}^+_{(1,1)} + \mathscr{D}_{(1,0)} + \mathscr{D}^+_{(0,0)};$$

the product of two trivectors decomposes into a scalar (the scalar product), a bivector, and a tensor equivalent to the set of harmonic polynomials of the second degree.

164. Decomposition of the product $\mathscr{D}^+_{(p/2,p/2)} \times \mathscr{D}_{(p'/2',q'/2)}$

The decomposition can be carried out in an analogous way. The irreducible parts of the product are $(p' \neq q')$ the $\mathscr{D}_{[(p+p'/2)-i,(p+q'/2)-j]}$ where i and j, independently of each other, take the values $(0 \leqslant i \leqslant p,\ 0 \leqslant i \leqslant p')$ and $(0 \leqslant j \leqslant p, 0 \leqslant j \leqslant q')$; but if $(p + p'/2) - i = (p + q'/2) - j$, it is necessary to include this tensor twice, once with a superfix $+$, and once with a superfix $-$.

For example, the product of a trivector and a spinor $(p = 1, p' = 1, q' = 0)$ leads to values 0 and 1 for i, and 0 for j; this gives $\mathscr{D}_{(1,\frac{1}{2})}$ and $\mathscr{D}_{(0,\frac{1}{2})}$. An equivalent decomposition can be obtained by starting from a vector and a spinor. In general the two products $\mathscr{D}^+_{(p/2,p/2)} \times \mathscr{D}_{(p'/2,q'/2)}$ and $\mathscr{D}^-_{(p/2,p/2)} \times \mathscr{D}_{(p'/2,q'/2)}$ are equivalent $(p' \neq q')$, even though the two products which both have the same second factor, have non-equivalent first factors.

In the example considered, the part of the decomposition $\mathscr{D}_{(1,\frac{1}{2})}$ has as generating polynomial

$$(-acx_{2'} + bcx_{1'} + adx_1 + bdx_2)(a\xi_1 + b\xi_2 + c\xi_0 + d\xi_{12}).$$

If we replace the vector x by $\partial/\partial x$, we obtain the irreducible tensor

$$\left(-ac\frac{\partial}{\partial x^{2'}} + bc\frac{\partial}{\partial x^{1'}} + ad\frac{\partial}{\partial x^1} + bd\frac{\partial}{\partial x^2}\right)\left(a\xi_1 + b\xi_2 + c\xi_0 + d\xi_{12}\right).$$

Each component of this tensor forms one of the irreducible parts of the derivative of a field of spinors. The spinor fields which make this tensor zero are given by

$$\xi_1 = \alpha x^{1'} + \beta x^2 + \gamma, \qquad \xi_2 = \alpha x^{2'} - \beta x^1 + \delta,$$

$$\xi_0 = \alpha' x^1 + \beta' x^2 + \gamma', \qquad \xi_{12} = \alpha' x^{2'} - \beta' x^{1'} + \delta',$$

where $\alpha, \beta, \gamma, \delta, \alpha', \beta', \gamma', \delta'$ are arbitrary constants. Any displacement or reversal whatsoever transforms a spinor field of this type into a spinor field of the same type.

165. Decomposition of the product $\mathscr{D}_{(p/2,\,q/2)} \times \mathscr{D}_{(p'/2,\,q'/2)}$

An analogous argument leads to the decomposition ($p' \neq q'$)

$$\mathscr{D}_{(p/2,q/2)} \times \mathscr{D}_{(p'/2,q'/2)} = \sum \mathscr{D}_{[(p+p'/2)-i,\,(q+q'/2)-j]} + \sum \mathscr{D}_{[(p+q'/2)-k,\,(q+p'/2)-h]},$$

where the sums extend over all values of i, j, k, h which satisfy

$$0 \leqslant i \leqslant p, 0 \leqslant i \leqslant p'; 0 \leqslant j \leqslant q, 0 \leqslant j \leqslant q';$$

$$0 \leqslant k \leqslant p, 0 \leqslant k \leqslant q', 0 \leqslant h \leqslant q, 0 \leqslant h \leqslant p'.$$

If we have

$$\frac{p+p'}{2} - i = \frac{q+q'}{2} - j \quad \text{or} \quad \frac{p+q'}{2} - k = \frac{q+p'}{2} - h$$

then we must replace the corresponding symbol $\mathscr{D}$ by $\mathscr{D}^+ + \mathscr{D}^-$.

For example, we have for the product of two spinors

$$\mathscr{D}_{(\frac{1}{2},0)} \times \mathscr{D}_{(\frac{1}{2},0)} = \mathscr{D}_{(1,0)} + \overset{+}{\mathscr{D}}_{(\frac{1}{2},\frac{1}{2})} + \overset{-}{\mathscr{D}}_{(\frac{1}{2},\frac{1}{2})} + \overset{+}{\mathscr{D}}_{(0,0)} + \overset{-}{\mathscr{D}}_{(0,0)};$$

the decomposition gives a bivector, a vector, a trivector, a scalar, and a four-vector, which agrees with the reduction obtained directly. In particular the vector and bivector are given by the generating polynomials

$$(a\xi_1 + b\xi_2)(c\xi'_0 + d\xi'_{12}) + (c\xi_0 + d\xi_{12})(a\xi'_1 + b\xi'_2),$$

$$(a\xi_1 + b\xi_2)(c\xi'_0 + d\xi'_{12}) - (c\xi_0 + d\xi_{12})(a\xi'_1 + b\xi'_2),$$

of which the former is equivalent to the polynomial $(a\xi_1 + b\xi_2)\,(c\xi_0 + d\xi_{12})$.

III. LINEAR REPRESENTATIONS OF THE GROUP OF ROTATIONS IN THE REAL EUCLIDEAN SPACE E_n

166. Introduction

E. Cartan has determined all irreducible linear representations of the group of rotations in real Euclidean space of n dimensions*. We also have the result, due to H. Weyl (Section 81, footnote‡), that the theorem of complete reducibility holds for all linear representations of this group. These results extend as they stand to the group of proper rotations in real pseudo-Euclidean spaces.

* See E. Cartan, "Les groupes projectifs qui ne laissent invariante aucune multiplicité plane", *Bull. Soc. Math. France*, **41**, 1913, 53–96. A totally different method is due to R. Brauer: "Über die Darstellung der Drehungsgruppe durch Gruppen linearer Substitutionen", *Inaugural-Dissertation*, Göttingen, 1925. For the group of complex rotations, see also R. Brauer: "Die stetigen Darstellungen der komplexen Orthogonalgruppe", *Sitzungsb. Akad. Berlin*, 1929, 3–15.

The method applied in Sections 82–84 enables us to deduce from this all the representations of the group of complex rotations.

We must distinguish between the case of odd n and even n.

167. Case of the space $E_{2\nu+1}$

We take as fundamental form

$$(x^0)^2 + x^1x^{1'} + x^2x^{2'} + \cdots + x^\nu x^{\nu'}.$$

Then any irreducible linear representation of the group of (proper) rotations can be obtained by starting from the expression

$$\xi_0^p(x^1)^{p_1}(x^{12})^{p_2}\ldots(x^{12\ldots(\nu-1)})^{p_{\nu-1}},$$

and applying to it the different rotations of the group. The exponents $p, p_1, p_2, \ldots, p_{\nu-1}$ are arbitrary positive or zero integers. For example the representation ($p = 2$, $p_1 = p_2 = \cdots = p_{\nu-1} = 0$) gives a ν-vector; in fact the ν-vector associated with a spinor ξ and obtained from the expression $\xi^T C \underset{(\nu)}{X} \xi$ contains the component $x_{1'2'\ldots\nu'} = \pm\xi_0^2$; it follows that on applying the various rotations to ξ_0^2 we obtain an irreducible tensor equivalent to that given by applying these rotations to the component $x^{12\ldots\nu}$ of a ν-vector, and in this latter case we obtain a ν-vector.

168. Case of the space $E_{2\nu}$

We still take as fundamental form

$$x^1x^{1'} + x^2x^{2'} + \cdots + x^\nu x^{\nu'}.$$

Any irreducible linear representation of the group of rotations can be obtained by starting from the expression

$$\xi_0^p\xi_\nu^{p_1}(x^1)^{p_2}\ldots(x^{12\ldots(\nu-2)})^{p_{\nu-1}},$$

and applying to it the different rotations of the group.

For $p = 1$, $p_1 = p_2 = \cdots = p_{\nu-1} = 0$ we obtain a semi-spinor with an even number of indices; for $p = 0, p_1 = 1, p_2 = \cdots = p_{\nu-1} = 0$ a semi-spinor with an odd number of indices. The tensors ($p = 2, p_1 = p_2 = \cdots = p_{\nu-1} = 0$) and ($p = 0, p_1 = 2, p_2 = \cdots = p_{\nu-1} = 0$) are semi-$\nu$-vectors of the first and second types; the tensor ($p = 1, p_1 = 1, p_2 = \cdots = p_{\nu-1} = 0$) is a $(\nu - 1)$-vector. To show the latter result it is only necessary to note that the expression

$$\varphi^T C \underset{(\nu-1)}{X} \psi$$

gives a $(\nu - 1)$-vector associated with two semi-spinors φ and ψ and that for this $(\nu - 1)$-vector we have

$$x_{1'2'\ldots(\nu-1)'} = \pm\xi_0\xi_\nu;$$

the tensor obtained by applying the various rotations to the product $\xi_0\xi_\nu$ is thus equivalent to that obtained by applying these rotations to $x^{12\ldots(\nu-1)}$, that is to a $(\nu - 1)$-vector.

169. Irreducible linear representations of the group of rotations and reversals

In the space $E_{2\nu+1}$, any reversal leaves invariant each of the ν irreducible fundamental tensors generated by the components $\xi_0, x^1, x^{12}, \ldots x^{12\ldots(\nu-1)}$; it follows that, given any linear representation whatsoever of the group of rotations and reversals which induces a reducible representation in the group of rotations, any reversal transforms one of these irreducible parts into another equivalent part. Thus, by the theorem of Section 88,

THEOREM. *Any irreducible representation of the group of (proper) rotations and reversals in $E_{2\nu+1}$ induces an irreducible representation in the group of rotations, and conversely to any irreducible representation of the group of rotations there correspond two non-equivalent irreducible representations of the group of rotations and reversals.*

In the case of the space $E_{2\nu}$, any reversal leaves invariant each of the irreducible tensors generated by $x^1, x^{12}, \ldots, x^{12\ldots(\nu-2)}$, but transforms a semi-spinor into a semi-spinor of a different type. There are thus two sorts of irreducible linear representations of the group of (proper) rotations and reversals in $E_{2\nu}$.

(i) Any irreducible tensor for which $p = p_1$ provides two non-equivalent irreducible linear representations of the group of rotations and reversals;

(ii) The set of all components of the two tensors $(p, p_1, p_2, \ldots, p_{\nu-1})$ and $(p_1, p, p_2, \ldots, p_{\nu-1})$ with $p \neq p_1$ provides one and only one irreducible representation of the group of rotations and reversals.

Note that, since the expression $\xi_0\xi_\nu$ generates a $(\nu - 1)$-vector, tensors of the first category, or rather half of these tensors, can be obtained by applying the various rotations to an expression of the form

$$(x^1)^{p_1}(x^{12})^{p_2}\ldots(x^{12\ldots\nu-1})^{p_{\nu-1}}.$$

As an example of an irreducible tensor of the second category we need only mention spinors.

We notice that there are fundamental differences between the case of space of even dimension and the case of space of odd dimension.

170. Particular case when $n = 8$

In the case $n = 8$ the first three fundamental irreducible representations of the group of (proper) rotations are semi-spinors of the first type, semi-spinors of the second type, and vectors. We have shown (Section 139) that it is possible to adjoin to the group of rotations five other families of operations of such a nature that all the operations of each family permute the three types of tensor under consideration in the same way; the permutation varies from one family to another. The components of the irreducible tensors of the total group formed by these six families under consideration are obtained by adding to the components of the tensor $(p, p_1, p_2, p_3,)$ the components of the tensors obtained by carrying out all possible permutations on the first three exponents

(p, p_1, p_2). We notice that the fourth fundamental tensor, namely bivectors, is invariant under all transformations of the total group. For example, it is easy to verify that the tensor $\xi_\alpha\xi'_\beta - \xi_\beta\xi'_\alpha$, where ξ_α and ξ'_α are components of two semi-spinors of the same type, is equivalent to a bivector, since in the bivector given by the expression $\varphi^T \underset{(2)}{CX}\varphi$, the component $x_{1'2'}$ is equal to $-\xi_0\xi'_{34} + \xi_{34}\xi'_0$; applying the various rotations to the expression $\xi_0\xi'_{34} - \xi_{34}\xi'_0$ thus gives a tensor equivalent to a bivector.

171. Note

The method of H. Weyl for the determination of irreducible linear representations of compact groups* enables us to find without difficulty the degree of each of the different irreducible linear representations indicated above. We cite just one example for the space E_8. The degree of the irreducible representation (p, p_1, p_2, p_3) is equal to

$$\frac{F(p+1, p_1+1, p_2+1, p_3+1)}{F(1, 1, 1, 1)}$$

where

$$F(x, x_1, x_2, x_3) = xx_1x_2x_3(x+x_3)(x_1+x_3)(x_2+x_3)(x+x_1+x_3) \times (x+x_2+x_3)(x_1+x_2+x_3)(x+x_1+x_2+x_3)(x+x_1+x_2+2x_3)$$

For example the degree of the representation (1, 1, 1, 0) is equal to 350.

* H. Weyl, "Theorie der Darstellung kontinuierlicher halbeinfacher Gruppen durch lineare Transformationen", *Math. Zeitschr.*, **23**, 1925, 271–309; **24**, 1925, 328–395.

CHAPTER IX

SPINORS AND DIRAC'S EQUATIONS IN RIEMANNIAN GEOMETRY

I. SPINOR FIELDS IN EUCLIDEAN GEOMETRY

172. Infinitesimal rotations acting on spinors

We have already shown (Section 19) that any infinitesimal rotation in a Euclidean space E_n forms a tensor equivalent to a bivector. We shall rederive this result by a different method which at the same time shows how this bivector operates on spinors.

Take first a *simple* rotation through an angle α given by a pair of reflections associated with two unit vectors making an angle $\alpha/2$ with each other. Let A_1 be the first of these unit vectors and let A_2 be the unit vector perpendicular to A_1 in the biplane of the simple rotation: the rotation under consideration will be represented by the matrix

$$\left(A_1 \cos\frac{\alpha}{2} + A_2 \sin\frac{\alpha}{2}\right)A_1 = \cos\frac{\alpha}{2} - \sin\frac{\alpha}{2} A_1 A_2.$$

If we assume α to be infinitely small, the principal part is $1 - \frac{1}{2}\alpha A_1 A_2$. The effect produced on a vector X will be given by

$$X' = (1 - \tfrac{1}{2}\alpha A_1 A_2)X(1 + \tfrac{1}{2}\alpha A_1 A_2) = X + \tfrac{1}{2}\alpha(XA_1A_2 - A_1A_2X);$$

the effect produced on a spinor ξ will be

$$\xi' = (1 - \tfrac{1}{2}\alpha A_1 A_2)\xi.$$

It follows that if we denote the matrix associated with the bivector $\alpha A_1 A_2$ by U, we have the infinitesimal transformation

$$\delta X = \tfrac{1}{2}(XU - UX), \tag{1}$$

$$\delta\xi = -\tfrac{1}{2}U\xi. \tag{2}$$

If, for example, we refer the space to n basis vectors $A_1, A_2, \ldots, A_n$ where

$$\tfrac{1}{2}(A_iA_j + A_jA_i) = g_{ij},$$

the infinitesimal rotation denoted by

$$U = \tfrac{1}{2}a^{ij}A_iA_j \qquad (a^{ij} = -a^{ji})$$

will give, when applied to a vector x^kA_k,

$$\begin{aligned}\delta x^k . A_k &= \tfrac{1}{4}a^{ij}x^k(A_kA_iA_j - A_iA_jA_k)\\ &= \tfrac{1}{8}a^{ij}x^k[A_k(A_iA_j - A_jA_i) - (A_iA_j - A_jA_i)A_k].\end{aligned}$$

The quantity in brackets is equal to

$$\begin{aligned}-A_i(A_jA_k + A_kA_j) + A_j(A_iA_k + A_kA_i) - (A_jA_k + A_kA_j)A_i + (A_iA_k + A_kA_i)A_j\\ = -4(g_{jk}A_i - g_{ik}A_j).\end{aligned}$$

We thus have

$$\delta x^k . A_k = -\tfrac{1}{2}a^{ij}x^k(g_{jk}A_i - g_{ij}A_j) = a^{ij}g_{ik}x^kA_j = a^j_kx^kA_j,$$

from which

$$\delta x^i = a^i_kx^k \quad \text{or} \quad \delta x^i = a^{ki}x_k \quad \text{or} \quad \delta x^i = a_{ki}x^k. \tag{3}$$

Formula (2) becomes

$$\delta\xi = -\tfrac{1}{4}a^{ij}A_iA_j\xi. \tag{2'}$$

For example, for $n = 3$, keeping to the notation of Chapter III, we can put

$$U = i(\alpha H_1 + \beta H_2 + VH_3)$$

$$\delta\xi = -\tfrac{1}{2}i(\alpha H_1 + \beta H_2 + \gamma H_3)\xi,$$

from which, taking into account the expressions (Section 55) for the matrices H_i,

$$\left.\begin{aligned}\delta\xi_0 &= -\tfrac{1}{2}i\gamma\xi_0 - \tfrac{1}{2}i(\alpha - \beta i)\xi_1,\\ \delta\xi_1 &= -\tfrac{1}{2}i(\alpha + \beta i)\xi_0 + \tfrac{1}{2}i\gamma\xi_1;\end{aligned}\right\} \tag{4}$$

in terms of the general notation given above, the coefficients α, β, γ are $\alpha = a_{23}$, $\beta = a_{31}, \gamma = a_{12}$.

173. Alternative intepretation

We can given a different interpretation to the results we have obtained. Imagine the components of a *fixed* vector or a *fixed* spinor referred successively to two Cartesian frames of reference (R) and (R′) of which the new frame (R′) is obtained from the old frame (R) by a given infinitesimal rotation. If we use the sign to denote components of the vector, or spinor, referred to the new frame, then the vector, or spinor, which when referred to the old frame has these new components, can be obtained from the given vector, or spinor, by the inverse of the given infinitesimal rotation. We have thus the

THEOREM. *Given a vector* X *or a spinor* ξ *referred successively to a frame of reference* (R) *and to a frame of reference* (R′) *which can be obtained from* (R) *by the infinitesimal rotation* U, *the infinitesimal variations undergone by the components of this vector or spinor are given by the formulae*

$$\delta X = \tfrac{1}{2}(UX - XU), \tag{5}$$

$$\delta \xi = \tfrac{1}{2}U\xi, \tag{6}$$

of which the first can be written

$$\delta x^i = -a^i_k x^k \quad \text{or} \quad \delta x^i = a^{ik} x_k \quad \text{or} \quad \delta x_i = a_{ik} x^k. \tag{7}$$

In particular the formulæ (4) become

$$\delta\xi_0 = \tfrac{1}{2}i\gamma\xi_0 + \tfrac{1}{2}i(\alpha - \beta i)\xi_1,$$

$$\delta\xi_1 = \tfrac{1}{2}i(\alpha + \beta_i)\xi_0 - \tfrac{1}{2}i\gamma\xi_1.$$

174. The absolute differential of a vector and of a spinor

Consider a field of vectors or a field of spinors in a given space, each vector (or spinor) being attached to a point in space. Imagine that there is at each point of space a Cartesian reference system so that, referred to these systems, the fundamental form at each point always has the same coefficients (i.e., the reference system is everywhere equal to itself in the sense of Euclidean geometry).

Let Ω be the bivector which represents the infinitesimal rotation which makes the reference frame (R) with origin M equipollent with the reference frame (R′) with the infinitely close origin M′; the components of this bivector will be denoted by ω^{ij}. Let x^i be the components of the field vector at M referred to (R) and $x^i + dx^i$ be the components of the field vector at M′ referred to (R′). To pass from the first to the second, we pass from the first to a vector at M′ having the same components x^i when referred to (R′); this can be done by using the rotation Ω which brings (R) to (R′); then by adding dx^i to x^i we can pass to the second vector.

The elementary geometric variation Dx^i which is undergone by the field vector is then, by Equation (3),

$$Dx^i = dx^i + \omega^i_k x^k \quad \text{or} \quad DX = dX + \tfrac{1}{2}(X\Omega - \Omega X); \tag{8}$$

for a spinor field, we have in the same way

$$D\xi = d\xi - \tfrac{1}{2}\Omega\xi; \tag{9}$$

dX or $d\xi$ represents the *relative variation*, $\tfrac{1}{2}(X\Omega - \Omega X)$ (or $-\tfrac{1}{2}\Omega\xi$) the *variation of transport*. DX and $D\xi$ are the *absolute differentials* of the vector or spinor.

175. Dirac's equations

Suppose space-time of special relativity to be referred to a Galilean frame of reference at each point of space and let

$$(\omega^1)^2 + (\omega^2)^2 + (\omega^3)^2 - c^2(\omega^4)^2$$

be the scalar square of the vector which joins the point M to the infinitesimally near point M′; $\omega^1, \omega^2, \omega^3, \omega^4$ are the contravariant components of the vector **MM′**. The matrix Ω which defines the infinitesimal rotation which brings the reference frame (R) with origin M into equipollence with the reference frame (R′) with origin M′ is (Section 149)

$$\Omega = \begin{pmatrix} \pi & 0 \\ 0 & \bar{\pi} \end{pmatrix}$$

with

$$\pi = \begin{pmatrix} -i(\omega^{12} - ic\omega^{34}) & -(\omega^{31} - ic\omega^{24}) + i(\omega^{23} - ic\omega^{14}) \\ \omega^{31} - ic\omega^{24} + i(\omega^{23} - ic\omega^{14}) & i(\omega^{12} - ic\omega^{34}) \end{pmatrix}. \quad (10)$$

If f defines a function of position (i.e., space-time), we write

$$df = d_1 f \,.\, \omega^1 + d_2 f \,.\, \omega^2 + d_3 f \,.\, \omega^3 + d_4 f \,.\, \omega^4;$$

when space is referred to the reference frame (R), the vector $\partial/\partial x$ must be replaced (Section 157) by

$$D = \begin{pmatrix} 0 & 0 & D_1 + iD_2 & D_3 - \frac{1}{c}D_4 \\ 0 & 0 & D_3 + \frac{1}{c}D_4 & -D_1 + iD_2 \\ D_1 - iD_2 & D_3 - \frac{1}{c}D_4 & 0 & 0 \\ D_3 + \frac{1}{c}D_4 & -D_1 - iD_2 & 0 & 0 \end{pmatrix}, \quad (11)$$

with the following meaning for the symbols D_i. We put

$$Df = D_1 f \,.\, \omega^1 + D_2 f \,.\, \omega^2 + D_3 f \,.\, \omega^3 + D_4 f \,.\, \omega^4, \quad (12)$$

and we have

$$\begin{aligned} D\xi_0 &= d\xi_0 + \frac{1}{2}\left(i\omega_{12} - \frac{1}{c}\omega_{34}\right)\xi_0 + \frac{1}{2}\left(\omega_{31} + \frac{i}{c}\omega_{24} - i\omega_{23} + \frac{1}{c}\omega_{14}\right)\xi_{12}, \\ D\xi_{12} &= d\xi_{12} - \frac{1}{2}\left(\omega_{31} + \frac{i}{c}\omega_{24} + i\omega_{23} - \frac{1}{c}\omega_{14}\right)\xi_0 - \frac{1}{2}\left(i\omega_{12} - \frac{1}{c}\omega_{34}\right)\xi_{12}, \\ D\xi_1 &= d\xi_1 - \frac{1}{2}\left(i\omega_{12} + \frac{1}{c}\omega_{34}\right)\xi_1 + \frac{1}{2}\left(\omega_{31} - \frac{i}{c}\omega_{24} + i\omega_{23} + \frac{1}{c}\omega_{14}\right)\xi_2, \\ D\xi_2 &= d\xi_2 - \frac{1}{2}\left(\omega_{31} - \frac{i}{c}\omega_{24} - i\omega_{23} - \frac{1}{c}\omega_{14}\right)\xi_1 + \frac{1}{2}\left(i\omega_{12} + \frac{1}{c}\omega_{34}\right)\xi_2. \end{aligned} \quad (13)$$

In equations (13) we have introduced the covariant components ω_{ij} which can be calculated from the contravariant components by the relations

$$\omega_{23} = \omega^{23}, \quad \omega_{31} = \omega^{31}, \quad \omega_{12} = \omega^{12}, \quad \omega_{14} = -c^2\omega^{14},$$
$$\omega_{24} = -c^2\omega^{24}, \quad \omega_{34} = -c^2\omega^{34}.$$

With this notation Dirac's equations can be written as the matrix relation

$$\left(\frac{h}{i}\mathrm{D} + \frac{e}{c}\mathrm{V} - m_0ck\right)\xi = 0 \tag{14}$$

whereas in Section 157, the matrix K is of the form

$$\begin{pmatrix} -1 & 0 \\ 0 & 1 \end{pmatrix},$$

each block representing a matrix of degree 2.

II. SPINOR FIELDS IN RIEMANNIAN GEOMETRY

176. Case of general relativity

There is no need to change the preceding formulæ provided we associate with each point in space-time a local Galilean reference frame which has as fundamental form

$$(\omega^1)^2 + (\omega^2)^2 + (\omega^3)^2 - c^2(\omega^4)^2,$$

and take the forms ω_{ij} as those which define the affine connection of space-time*.

Let us take as an example the $(\mathrm{d}s)^2$ of Schwarzschild (with a change of sign); this corresponds to the forms

$$\omega^1 = r\,\mathrm{d}\theta, \qquad \omega^2 = r\sin\theta\,\mathrm{d}\varphi, \qquad \omega^3 = \frac{\mathrm{d}r}{\sqrt{1-\dfrac{2m}{r}}}, \qquad \omega^4 = \sqrt{1-\frac{2m}{r}}\,\mathrm{d}t.$$

The only non-zero forms ω_{ij} are

$$\omega_{12} = \cos\theta\,\mathrm{d}\varphi, \qquad \omega_{31} = \sqrt{1-\frac{2m}{r}}\,\mathrm{d}\theta, \qquad \omega_{32} = \sqrt{1-\frac{2m}{r}}\sin\theta\,\mathrm{d}\varphi,$$

$$\omega_{34} = -\frac{c^2m}{r^2}\,\mathrm{d}t.$$

The matrix D is thus of the form

$$\begin{pmatrix} 0 & \Delta \\ \bar{\Delta} & 0 \end{pmatrix},$$

where Δ and $\bar{\Delta}$ are two complex conjugate matrices, and where we have

$$\Delta = \begin{pmatrix} a_{11} & a_{12} \\ a_{21} & a_{22} \end{pmatrix}, \tag{15}$$

* The proof given in Section 158 of the theorem which implies the result that the divergence of the current vector is zero, can be repeated without modification here.

with

$$\left.\begin{aligned}
a_{11} &= \frac{1}{r}\frac{\partial}{\partial\theta} + \frac{i}{r\sin\theta}\frac{\partial}{\partial\varphi} + \frac{\cot\theta}{2r}, \\
a_{12} &= \sqrt{1-\frac{2m}{r}}\frac{\partial}{\partial r} - \frac{1}{c\sqrt{1-\frac{2m}{r}}}\frac{\partial}{\partial t} + \frac{2r-3m}{2r^2\sqrt{1-\frac{2m}{r}}}, \\
a_{21} &= \sqrt{1-\frac{2m}{r}}\frac{\partial}{\partial r} + \frac{1}{c\sqrt{1-\frac{2m}{r}}}\frac{\partial}{\partial t} + \frac{2r-3m}{2r^2\sqrt{1-\frac{2m}{r}}}, \\
a_{22} &= -\frac{1}{r}\frac{\partial}{\partial\theta} + \frac{i}{r\sin\theta}\frac{\partial}{\partial\varphi} - \frac{\cot\theta}{2r};
\end{aligned}\right\} \tag{16}$$

$\bar{\Delta}$ can be deduced from Δ by changing each i into $-i$.

177. Case of any Cartesian reference frame whatsoever

In the preceding discussion we have assumed that the Riemannian space is referred to a particular family of Cartesian frames of reference; and it is necessary that from the metrical point of view these should be equal to one another. This is the point of view which has been adopted by most authors who have sought to extend Dirac's equations to general relativity*. But these authors do not regard spinors as well-defined geometric entities; V. Fock deduces the law of transformation of a spinor from the law of transformation of the vector generated by the spinor and its conjugate; this introduces an indeterminacy which he regards as giving rise to an electromagnetic field. Other physicists, not wishing to employ local Galilean reference frames, have sought to generalise Dirac's equations by using the classical technique of Riemannian geometry†; this technique rests on the use of Cartesian reference frames which depend on the choice of co-ordinates; these reference frames can, from the metrical point of view, have any fundamental forms whatsoever. We shall see that if we adopt this point of view and wish to continue to regard spinors as well-defined geometric entities, which behave as tensors in the most general sense of that term, then the generalisation of Dirac's equations will become impossible‡.

The point at issue is whether spinors retain their tensorial character under all linear transformations of n variables. In reality the problem is slightly less restrictive, since certain Euclidean tensors of a certain degree r,

---

* See for example E. Schrödinger, "Diracsches Elektron im Schwerfeld", *Sitzungsb. Akad. f. Physik*, **57**, 1929, p. 261–277 and also H. Weyl. "Elektron und Gravitation", *Zeitsch. f. Physik*, **56**, 1929, 330–352, who takes a different point of view.

† See for example E. Schrödinger, "Diracsches Elektron im Schwerfeld", *Sitzungsb. Akad. Berlin*, 1932, 105.

‡ Certain physicists regard spinors as entities which are, in a sense, unaffected by the rotations which classical geometric entities (vectors etc.) can undergo, and of which the components in a given reference system are susceptible of undergoing linear transformations which are in a sense autonomous. See for example L. Infeld and B. L. van der Waerden, "Die Wellengleichungen des Elektrons in der allgemeinen Relativitätstheorie", *Sitzungsb. Akad. Berlin*, 1933, 380.

which are not affine tensors, can nevertheless be analytically defined with respect to any Cartesian reference frame whatsoever, by the components of an affine tensor of degree R (greater than r) such that, when referred to a rectangular frame of reference, $R - r$ of these components vanish, the r other components being those of the given Euclidean tensor. Thus, for $n = 4$, a semi-bivector can be represented by the components $p_{ij} + \sqrt{g}p^{kh}$ ($ijkh$ is an even permutation) of a bivector. We thus require to know whether it is possible, in the affine space of $n = 2\nu + 1$ or 2ν dimensions, to find an affine tensor with N components such that, referred to a Cartesian frame of reference, $N - 2^\nu$ of these components vanish, and the other 2^ν components behave as the components of a spinor under a change of rectangular frame of reference. We shall see that this is impossible.

By using the passage from real to complex, we can assume that we are dealing with the complex domain. Suppose that the fundamental form has been reduced to a sum of squares

$$F \equiv x_1^2 + x_2^2 + \cdots + x_n^2.$$

We shall have a linear representation of the group of all linear substitutions; this will provide, in particular, a linear representation of the group

$$x'_1 = \alpha x_1 + \beta x_2, \quad x'_2 = \gamma x_1 + \delta x_2, \quad x'_3 = x_3, \ldots, x'_n = x^n \quad (\alpha\delta - \beta\gamma = 1). \quad (17)$$

This representation will be many valued, since under the Euclidean rotation

$$x'_1 = x_1 \cos\theta - x_2 \sin\theta, \quad x'_2 = x_1 \sin\theta + x_2 \cos\theta, \quad x'_3 = x_3, \ldots, x'_n = x_n,$$

where θ is real and varies in a continuous manner from 0 to 2π, the spinor undergoes a linear substitution which passes in a continuous manner from the identity substitution $\xi' = \xi$ to the substitution $\xi' = -\xi$. *We shall thus have a many-valued representation of the group* (17), and we have seen (Section 85) that this is impossible.

We could take, without making any change in the proof, instead of the group (17), the unimodular unitary group in the variables x_1 and x_2; we have given (in Section 85), a topological argument to show the non-existence of multivalued linear representations of this group.

We thus arrive at the following fundamental theorem.

THEOREM. *With the geometric sense we have given to the word "spinor" it is impossible to introduce fields of spinors into the classical Riemannian technique*; that is, *having chosen an arbitrary system of co-ordinates x^i for the space, it is impossible to represent a spinor by any finite number* N *whatsoever, of components u_α such that the u_α have covariant derivatives of the form*

$$u_{\alpha,i} = \frac{\partial u_\alpha}{\partial x^i} + \Lambda^\beta_{\alpha i} u_\beta,$$

*where the $\Lambda^\beta_{\alpha i}$ are determinate functions of x^h**.

* It is clear that this impossibility provides an explanation of the point of view of L. Infeld and van der Waerden (see the preceding footnote), which is however geometrically and even physically so startling.

BIBLIOGRAPHY

Cartan's theory of spinors was developed by R. Brauer and H. Weyl in:

"Spinors in *n* Dimensions", *American Journal of Mathematics*, **57**, 425, (1935).

Spinor theory is treated as part of the more general theory of group representations in the books :

H. Weyl, *The Classical Groups*, Princeton University Press, 1946.

C. Chevalley, *Theory of Lie Groups*, Princeton University Press, 1946.

H. Boener, *Representations of Groups*, North Holland, 1963.

The development of the algebraic aspects of spinors is presented in the advanced monograph:

C. Chevalley, *The Algebraic Theory of Spinors*, Columbia University Press, 1954.

The first application of spinors to physics was by W. Pauli, who introduced his famous spin matrices:

W. Pauli, "Zur Quantenmechanik des magnetischen Elektrons", *Z. Physik*, **43**, 601 (1927).

The fully relativistic theory of the electron spin was discovered by Dirac, who showed the connection between spinors and the Lorentz group:

P. A. M. Dirac, *Proc. Roy. Soc.*, **A117**, 610 (1928).

The connection with the Lorentz group was treated mathematically by:

B. L. van der Waerden, Nachr. Wiss. Gottingen, *Math.-Physik*, 100 (1929),

and

B. L. van der Waerden, *Die gruppentheoretische Methode in der Quantenmechanik*, Springer, Berlin, (1932).

A vast literature on relativistic wave equations developed between the wars and soon after. A useful handbook for spinor equations, including an extensive bibliography, is:

E. M. Corson, *Introduction to Tensors, Spinors and Relativistic Wave Equations*, Blackie, Glasgow, (1953).

Undoubtedly, the most important single paper on the subject is that of Fierz:

M. Fierz: "Uber die Relativistische Theorie kraftefreier Teilchen mit beliebigen Spin" *Helv. Phys. Acta*, **12**, 3 (1938).

The connection between spinor wave equations and the representations of the inhomogeneous Lorentz group was first made clear by Bargmann and Wigner:

V. Bargmann and E. P. Wigner, "Group-theoretical Discussion of Relativistic Wave Equations", *Proc. Nat. Acad. Sci.*, **34**, 211 (1948).

This point of view has been developed by H. Joos, *Fortschritte der Physik*, **10**, 65 (1962) and especially:

S. Weinberg, "Feynman Rules for Any Spin", *Phys. Rev.*, **133**, B1318 (1964) and **139**, B597 (1965).

D. Pursey has developed a technique for constructing all spinor equations associated with a given representation of the inhomogeneous Lorentz group:

D. L. Pursey, *Annals of Physics* (to appear).

A summary of the applications of spinors to the homogeneous and inhomogeneous Lorentz group is to be found in the book T. Kahan (editor)

Theory of Groups in Classical and Quantum Theory, Oliver and Boyd, (London), 1965, Vol. I (translated from *Théorie des groupes en physique classique et quantique*, Dunod).

The proof of the famous PCT theorem for any spinor field is given in Pauli's contribution to:

Niels Bohr and the Development of Physics, Pergamon, Oxford, (1955).

For a more modern treatment, see:

R. F. Streater and A. S. Wightman, *PCT, Spin and Statistics, and All That*, Benjamin, (New York), 1964.

Spinor Theory can be applied to atomic physics; the classic work on this subject is:

E. P. Wigner, *Group Theory and its Application to the Quantum Mechanics of Atomic Spectra*, Academic Press, New York, (1959).

Spinors may be applied to general relativity; see:

O. Veblen and J. von Neumann, "Geometry of Complex Domains", mimeographed notes, issued by the Institute for Advanced Study, Princeton, 1936 (reissued 1955).

R. Penrose, *Annals of Physics*, **10**, 171 (1960).

Attempts have been made to classify elementary particles using spinor representations of groups. For example:

B. d'Espagnat and J. Prentki, *Nuclear Physics*, **1**, 33 (1956).

A more complete bibliography can be found in the review article of:

W. C. Parke and H. Jehle in *Lectures in Theoretical Physics, Boulder, 1964*, Vol.VIIa, University of Colorado Press, 1965.

SUBJECT INDEX

Reference numbers indicate in turn
chapter
subchapter
paragraph
page

A CATALOG OF SELECTED

DOVER BOOKS

IN SCIENCE AND MATHEMATICS

Astronomy

BURNHAM'S CELESTIAL HANDBOOK, Robert Burnham, Jr. Thorough guide to the stars beyond our solar system. Exhaustive treatment. Alphabetical by constellation: Andromeda to Cetus in Vol. 1; Chamaeleon to Orion in Vol. 2; and Pavo to Vulpecula in Vol. 3. Hundreds of illustrations. Index in Vol. 3. 2,000pp. 6⅛ x 9¼.
Vol. I: 0-486-23567-X
Vol. II: 0-486-23568-8
Vol. III: 0-486-23673-0

EXPLORING THE MOON THROUGH BINOCULARS AND SMALL TELESCOPES, Ernest H. Cherrington, Jr. Informative, profusely illustrated guide to locating and identifying craters, rills, seas, mountains, other lunar features. Newly revised and updated with special section of new photos. Over 100 photos and diagrams. 240pp. 8¼ x 11. 0-486-24491-1

THE EXTRATERRESTRIAL LIFE DEBATE, 1750–1900, Michael J. Crowe. First detailed, scholarly study in English of the many ideas that developed from 1750 to 1900 regarding the existence of intelligent extraterrestrial life. Examines ideas of Kant, Herschel, Voltaire, Percival Lowell, many other scientists and thinkers. 16 illustrations. 704pp. 5⅜ x 8½. 0-486-40675-X

THEORIES OF THE WORLD FROM ANTIQUITY TO THE COPERNICAN REVOLUTION, Michael J. Crowe. Newly revised edition of an accessible, enlightening book recreates the change from an earth-centered to a sun-centered conception of the solar system. 242pp. 5⅜ x 8½. 0-486-41444-2

A HISTORY OF ASTRONOMY, A. Pannekoek. Well-balanced, carefully reasoned study covers such topics as Ptolemaic theory, work of Copernicus, Kepler, Newton, Eddington's work on stars, much more. Illustrated. References. 521pp. 5⅜ x 8½.
0-486-65994-1

A COMPLETE MANUAL OF AMATEUR ASTRONOMY: TOOLS AND TECHNIQUES FOR ASTRONOMICAL OBSERVATIONS, P. Clay Sherrod with Thomas L. Koed. Concise, highly readable book discusses: selecting, setting up and maintaining a telescope; amateur studies of the sun; lunar topography and occultations; observations of Mars, Jupiter, Saturn, the minor planets and the stars; an introduction to photoelectric photometry; more. 1981 ed. 124 figures. 25 halftones. 37 tables. 335pp. 6½ x 9¼. 0-486-40675-X

AMATEUR ASTRONOMER'S HANDBOOK, J. B. Sidgwick. Timeless, comprehensive coverage of telescopes, mirrors, lenses, mountings, telescope drives, micrometers, spectroscopes, more. 189 illustrations. 576pp. 5⅝ x 8¼. (Available in U.S. only.)
0-486-24034-7

STARS AND RELATIVITY, Ya. B. Zel'dovich and I. D. Novikov. Vol. 1 of *Relativistic Astrophysics* by famed Russian scientists. General relativity, properties of matter under astrophysical conditions, stars, and stellar systems. Deep physical insights, clear presentation. 1971 edition. References. 544pp. 5⅝ x 8¼. 0-486-69424-0

Chemistry

THE SCEPTICAL CHYMIST: THE CLASSIC 1661 TEXT, Robert Boyle. Boyle defines the term "element," asserting that all natural phenomena can be explained by the motion and organization of primary particles. 1911 ed. viii+232pp. 5⅜ x 8½. 0-486-42825-7

RADIOACTIVE SUBSTANCES, Marie Curie. Here is the celebrated scientist's doctoral thesis, the prelude to her receipt of the 1903 Nobel Prize. Curie discusses establishing atomic character of radioactivity found in compounds of uranium and thorium; extraction from pitchblende of polonium and radium; isolation of pure radium chloride; determination of atomic weight of radium; plus electric, photographic, luminous, heat, color effects of radioactivity. ii+94pp. 5⅜ x 8½. 0-486-42550-9

CHEMICAL MAGIC, Leonard A. Ford. Second Edition, Revised by E. Winston Grundmeier. Over 100 unusual stunts demonstrating cold fire, dust explosions, much more. Text explains scientific principles and stresses safety precautions. 128pp. 5⅜ x 8½. 0-486-67628-5

THE DEVELOPMENT OF MODERN CHEMISTRY, Aaron J. Ihde. Authoritative history of chemistry from ancient Greek theory to 20th-century innovation. Covers major chemists and their discoveries. 209 illustrations. 14 tables. Bibliographies. Indices. Appendices. 851pp. 5⅜ x 8½. 0-486-64235-6

CATALYSIS IN CHEMISTRY AND ENZYMOLOGY, William P. Jencks. Exceptionally clear coverage of mechanisms for catalysis, forces in aqueous solution, carbonyl- and acyl-group reactions, practical kinetics, more. 864pp. 5⅜ x 8½. 0-486-65460-5

ELEMENTS OF CHEMISTRY, Antoine Lavoisier. Monumental classic by founder of modern chemistry in remarkable reprint of rare 1790 Kerr translation. A must for every student of chemistry or the history of science. 539pp. 5⅜ x 8½. 0-486-64624-6

THE HISTORICAL BACKGROUND OF CHEMISTRY, Henry M. Leicester. Evolution of ideas, not individual biography. Concentrates on formulation of a coherent set of chemical laws. 260pp. 5⅜ x 8½. 0-486-61053-5

A SHORT HISTORY OF CHEMISTRY, J. R. Partington. Classic exposition explores origins of chemistry, alchemy, early medical chemistry, nature of atmosphere, theory of valency, laws and structure of atomic theory, much more. 428pp. 5⅜ x 8½. (Available in U.S. only.) 0-486-65977-1

GENERAL CHEMISTRY, Linus Pauling. Revised 3rd edition of classic first-year text by Nobel laureate. Atomic and molecular structure, quantum mechanics, statistical mechanics, thermodynamics correlated with descriptive chemistry. Problems. 992pp. 5⅜ x 8½. 0-486-65622-5

FROM ALCHEMY TO CHEMISTRY, John Read. Broad, humanistic treatment focuses on great figures of chemistry and ideas that revolutionized the science. 50 illustrations. 240pp. 5⅜ x 8½. 0-486-28690-8

Engineering

DE RE METALLICA, Georgius Agricola. The famous Hoover translation of greatest treatise on technological chemistry, engineering, geology, mining of early modern times (1556). All 289 original woodcuts. 638pp. 6¾ x 11. 0-486-60006-8

FUNDAMENTALS OF ASTRODYNAMICS, Roger Bate et al. Modern approach developed by U.S. Air Force Academy. Designed as a first course. Problems, exercises. Numerous illustrations. 455pp. 5⅜ x 8½. 0-486-60061-0

DYNAMICS OF FLUIDS IN POROUS MEDIA, Jacob Bear. For advanced students of ground water hydrology, soil mechanics and physics, drainage and irrigation engineering and more. 335 illustrations. Exercises, with answers. 784pp. 6⅛ x 9¼. 0-486-65675-6

THEORY OF VISCOELASTICITY (Second Edition), Richard M. Christensen. Complete consistent description of the linear theory of the viscoelastic behavior of materials. Problem-solving techniques discussed. 1982 edition. 29 figures. xiv+364pp. 6⅛ x 9¼. 0-486-42880-X

MECHANICS, J. P. Den Hartog. A classic introductory text or refresher. Hundreds of applications and design problems illuminate fundamentals of trusses, loaded beams and cables, etc. 334 answered problems. 462pp. 5⅜ x 8½. 0-486-60754-2

MECHANICAL VIBRATIONS, J. P. Den Hartog. Classic textbook offers lucid explanations and illustrative models, applying theories of vibrations to a variety of practical industrial engineering problems. Numerous figures. 233 problems, solutions. Appendix. Index. Preface. 436pp. 5⅜ x 8½. 0-486-64785-4

STRENGTH OF MATERIALS, J. P. Den Hartog. Full, clear treatment of basic material (tension, torsion, bending, etc.) plus advanced material on engineering methods, applications. 350 answered problems. 323pp. 5⅜ x 8½. 0-486-60755-0

A HISTORY OF MECHANICS, René Dugas. Monumental study of mechanical principles from antiquity to quantum mechanics. Contributions of ancient Greeks, Galileo, Leonardo, Kepler, Lagrange, many others. 671pp. 5⅜ x 8½. 0-486-65632-2

STABILITY THEORY AND ITS APPLICATIONS TO STRUCTURAL MECHANICS, Clive L. Dym. Self-contained text focuses on Koiter postbuckling analyses, with mathematical notions of stability of motion. Basing minimum energy principles for static stability upon dynamic concepts of stability of motion, it develops asymptotic buckling and postbuckling analyses from potential energy considerations, with applications to columns, plates, and arches. 1974 ed. 208pp. 5⅜ x 8½. 0-486-42541-X

METAL FATIGUE, N. E. Frost, K. J. Marsh, and L. P. Pook. Definitive, clearly written, and well-illustrated volume addresses all aspects of the subject, from the historical development of understanding metal fatigue to vital concepts of the cyclic stress that causes a crack to grow. Includes 7 appendixes. 544pp. 5⅜ x 8½. 0-486-40927-9

ROCKETS, Robert Goddard. Two of the most significant publications in the history of rocketry and jet propulsion: "A Method of Reaching Extreme Altitudes" (1919) and "Liquid Propellant Rocket Development" (1936). 128pp. 5⅜ x 8½. 0-486-42537-1

STATISTICAL MECHANICS: PRINCIPLES AND APPLICATIONS, Terrell L. Hill. Standard text covers fundamentals of statistical mechanics, applications to fluctuation theory, imperfect gases, distribution functions, more. 448pp. 5⅜ x 8½. 0-486-65390-0

ENGINEERING AND TECHNOLOGY 1650–1750: ILLUSTRATIONS AND TEXTS FROM ORIGINAL SOURCES, Martin Jensen. Highly readable text with more than 200 contemporary drawings and detailed engravings of engineering projects dealing with surveying, leveling, materials, hand tools, lifting equipment, transport and erection, piling, bailing, water supply, hydraulic engineering, and more. Among the specific projects outlined-transporting a 50-ton stone to the Louvre, erecting an obelisk, building timber locks, and dredging canals. 207pp. 8⅜ x 11¼. 0-486-42232-1

THE VARIATIONAL PRINCIPLES OF MECHANICS, Cornelius Lanczos. Graduate level coverage of calculus of variations, equations of motion, relativistic mechanics, more. First inexpensive paperbound edition of classic treatise. Index. Bibliography. 418pp. 5⅜ x 8½. 0-486-65067-7

PROTECTION OF ELECTRONIC CIRCUITS FROM OVERVOLTAGES, Ronald B. Standler. Five-part treatment presents practical rules and strategies for circuits designed to protect electronic systems from damage by transient overvoltages. 1989 ed. xxiv+434pp. 6⅛ x 9¼. 0-486-42552-5

ROTARY WING AERODYNAMICS, W. Z. Stepniewski. Clear, concise text covers aerodynamic phenomena of the rotor and offers guidelines for helicopter performance evaluation. Originally prepared for NASA. 537 figures. 640pp. 6⅛ x 9¼. 0-486-64647-5

INTRODUCTION TO SPACE DYNAMICS, William Tyrrell Thomson. Comprehensive, classic introduction to space-flight engineering for advanced undergraduate and graduate students. Includes vector algebra, kinematics, transformation of coordinates. Bibliography. Index. 352pp. 5⅜ x 8½. 0-486-65113-4

HISTORY OF STRENGTH OF MATERIALS, Stephen P. Timoshenko. Excellent historical survey of the strength of materials with many references to the theories of elasticity and structure. 245 figures. 452pp. 5⅜ x 8½. 0-486-61187-6

ANALYTICAL FRACTURE MECHANICS, David J. Unger. Self-contained text supplements standard fracture mechanics texts by focusing on analytical methods for determining crack-tip stress and strain fields. 336pp. 6⅛ x 9¼. 0-486-41737-9

STATISTICAL MECHANICS OF ELASTICITY, J. H. Weiner. Advanced, self-contained treatment illustrates general principles and elastic behavior of solids. Part 1, based on classical mechanics, studies thermoelastic behavior of crystalline and polymeric solids. Part 2, based on quantum mechanics, focuses on interatomic force laws, behavior of solids, and thermally activated processes. For students of physics and chemistry and for polymer physicists. 1983 ed. 96 figures. 496pp. 5⅜ x 8½. 0-486-42260-7

Mathematics

FUNCTIONAL ANALYSIS (Second Corrected Edition), George Bachman and Lawrence Narici. Excellent treatment of subject geared toward students with background in linear algebra, advanced calculus, physics and engineering. Text covers introduction to inner-product spaces, normed, metric spaces, and topological spaces; complete orthonormal sets, the Hahn-Banach Theorem and its consequences, and many other related subjects. 1966 ed. 544pp. 6⅛ x 9¼. 0-486-40251-7

ASYMPTOTIC EXPANSIONS OF INTEGRALS, Norman Bleistein & Richard A. Handelsman. Best introduction to important field with applications in a variety of scientific disciplines. New preface. Problems. Diagrams. Tables. Bibliography. Index. 448pp. 5⅜ x 8½. 0-486-65082-0

VECTOR AND TENSOR ANALYSIS WITH APPLICATIONS, A. I. Borisenko and I. E. Tarapov. Concise introduction. Worked-out problems, solutions, exercises. 257pp. 5⅝ x 8¼. 0-486-63833-2

AN INTRODUCTION TO ORDINARY DIFFERENTIAL EQUATIONS, Earl A. Coddington. A thorough and systematic first course in elementary differential equations for undergraduates in mathematics and science, with many exercises and problems (with answers). Index. 304pp. 5⅜ x 8½. 0-486-65942-9

FOURIER SERIES AND ORTHOGONAL FUNCTIONS, Harry F. Davis. An incisive text combining theory and practical example to introduce Fourier series, orthogonal functions and applications of the Fourier method to boundary-value problems. 570 exercises. Answers and notes. 416pp. 5⅜ x 8½. 0-486-65973-9

COMPUTABILITY AND UNSOLVABILITY, Martin Davis. Classic graduate-level introduction to theory of computability, usually referred to as theory of recurrent functions. New preface and appendix. 288pp. 5⅜ x 8½. 0-486-61471-9

ASYMPTOTIC METHODS IN ANALYSIS, N. G. de Bruijn. An inexpensive, comprehensive guide to asymptotic methods–the pioneering work that teaches by explaining worked examples in detail. Index. 224pp. 5⅜ x 8½ 0-486-64221-6

APPLIED COMPLEX VARIABLES, John W. Dettman. Step-by-step coverage of fundamentals of analytic function theory–plus lucid exposition of five important applications: Potential Theory; Ordinary Differential Equations; Fourier Transforms; Laplace Transforms; Asymptotic Expansions. 66 figures. Exercises at chapter ends. 512pp. 5⅜ x 8½. 0-486-64670-X

INTRODUCTION TO LINEAR ALGEBRA AND DIFFERENTIAL EQUATIONS, John W. Dettman. Excellent text covers complex numbers, determinants, orthonormal bases, Laplace transforms, much more. Exercises with solutions. Undergraduate level. 416pp. 5⅜ x 8½. 0-486-65191-6

RIEMANN'S ZETA FUNCTION, H. M. Edwards. Superb, high-level study of landmark 1859 publication entitled "On the Number of Primes Less Than a Given Magnitude" traces developments in mathematical theory that it inspired. xiv+315pp. 5⅜ x 8½. 0-486-41740-9

Math–Decision Theory, Statistics, Probability

ELEMENTARY DECISION THEORY, Herman Chernoff and Lincoln E. Moses. Clear introduction to statistics and statistical theory covers data processing, probability and random variables, testing hypotheses, much more. Exercises. 364pp. 5⅜ x 8½. 0-486-65218-1

STATISTICS MANUAL, Edwin L. Crow et al. Comprehensive, practical collection of classical and modern methods prepared by U.S. Naval Ordnance Test Station. Stress on use. Basics of statistics assumed. 288pp. 5⅜ x 8½. 0-486-60599-X

SOME THEORY OF SAMPLING, William Edwards Deming. Analysis of the problems, theory and design of sampling techniques for social scientists, industrial managers and others who find statistics important at work. 61 tables. 90 figures. xvii +602pp. 5⅜ x 8½. 0-486-64684-X

LINEAR PROGRAMMING AND ECONOMIC ANALYSIS, Robert Dorfman, Paul A. Samuelson and Robert M. Solow. First comprehensive treatment of linear programming in standard economic analysis. Game theory, modern welfare economics, Leontief input-output, more. 525pp. 5⅜ x 8½. 0-486-65491-5

PROBABILITY: AN INTRODUCTION, Samuel Goldberg. Excellent basic text covers set theory, probability theory for finite sample spaces, binomial theorem, much more. 360 problems. Bibliographies. 322pp. 5⅜ x 8½. 0-486-65252-1

GAMES AND DECISIONS: INTRODUCTION AND CRITICAL SURVEY, R. Duncan Luce and Howard Raiffa. Superb nontechnical introduction to game theory, primarily applied to social sciences. Utility theory, zero-sum games, n-person games, decision-making, much more. Bibliography. 509pp. 5⅜ x 8½. 0-486-65943-7

INTRODUCTION TO THE THEORY OF GAMES, J. C. C. McKinsey. This comprehensive overview of the mathematical theory of games illustrates applications to situations involving conflicts of interest, including economic, social, political, and military contexts. Appropriate for advanced undergraduate and graduate courses; advanced calculus a prerequisite. 1952 ed. x+372pp. 5⅜ x 8½. 0-486-42811-7

FIFTY CHALLENGING PROBLEMS IN PROBABILITY WITH SOLUTIONS, Frederick Mosteller. Remarkable puzzlers, graded in difficulty, illustrate elementary and advanced aspects of probability. Detailed solutions. 88pp. 5⅜ x 8½. 65355-2

PROBABILITY THEORY: A CONCISE COURSE, Y. A. Rozanov. Highly readable, self-contained introduction covers combination of events, dependent events, Bernoulli trials, etc. 148pp. 5⅜ x 8¼. 0-486-63544-9

STATISTICAL METHOD FROM THE VIEWPOINT OF QUALITY CONTROL, Walter A. Shewhart. Important text explains regulation of variables, uses of statistical control to achieve quality control in industry, agriculture, other areas. 192pp. 5⅜ x 8½. 0-486-65232-7

Math–Geometry and Topology

ELEMENTARY CONCEPTS OF TOPOLOGY, Paul Alexandroff. Elegant, intuitive approach to topology from set-theoretic topology to Betti groups; how concepts of topology are useful in math and physics. 25 figures. 57pp. 5⅜ x 8½. 0-486-60747-X

COMBINATORIAL TOPOLOGY, P. S. Alexandrov. Clearly written, well-organized, three-part text begins by dealing with certain classic problems without using the formal techniques of homology theory and advances to the central concept, the Betti groups. Numerous detailed examples. 654pp. 5⅜ x 8½. 0-486-40179-0

EXPERIMENTS IN TOPOLOGY, Stephen Barr. Classic, lively explanation of one of the byways of mathematics. Klein bottles, Moebius strips, projective planes, map coloring, problem of the Koenigsberg bridges, much more, described with clarity and wit. 43 figures. 210pp. 5⅜ x 8½. 0-486-25933-1

THE GEOMETRY OF RENÉ DESCARTES, René Descartes. The great work founded analytical geometry. Original French text, Descartes's own diagrams, together with definitive Smith-Latham translation. 244pp. 5⅜ x 8½. 0-486-60068-8

EUCLIDEAN GEOMETRY AND TRANSFORMATIONS, Clayton W. Dodge. This introduction to Euclidean geometry emphasizes transformations, particularly isometries and similarities. Suitable for undergraduate courses, it includes numerous examples, many with detailed answers. 1972 ed. viii+296pp. 6⅛ x 9¼. 0-486-43476-1

PRACTICAL CONIC SECTIONS: THE GEOMETRIC PROPERTIES OF ELLIPSES, PARABOLAS AND HYPERBOLAS, J. W. Downs. This text shows how to create ellipses, parabolas, and hyperbolas. It also presents historical background on their ancient origins and describes the reflective properties and roles of curves in design applications. 1993 ed. 98 figures. xii+100pp. 6½ x 9¼. 0-486-42876-1

THE THIRTEEN BOOKS OF EUCLID'S ELEMENTS, translated with introduction and commentary by Sir Thomas L. Heath. Definitive edition. Textual and linguistic notes, mathematical analysis. 2,500 years of critical commentary. Unabridged. 1,414pp. 5⅜ x 8½. Three-vol. set.
Vol. I: 0-486-60088-2 Vol. II: 0-486-60089-0 Vol. III: 0-486-60090-4

SPACE AND GEOMETRY: IN THE LIGHT OF PHYSIOLOGICAL, PSYCHOLOGICAL AND PHYSICAL INQUIRY, Ernst Mach. Three essays by an eminent philosopher and scientist explore the nature, origin, and development of our concepts of space, with a distinctness and precision suitable for undergraduate students and other readers. 1906 ed. vi+148pp. 5⅜ x 8½. 0-486-43909-7

GEOMETRY OF COMPLEX NUMBERS, Hans Schwerdtfeger. Illuminating, widely praised book on analytic geometry of circles, the Moebius transformation, and two-dimensional non-Euclidean geometries. 200pp. 5⅝ x 8¼. 0-486-63830-8

DIFFERENTIAL GEOMETRY, Heinrich W. Guggenheimer. Local differential geometry as an application of advanced calculus and linear algebra. Curvature, transformation groups, surfaces, more. Exercises. 62 figures. 378pp. 5⅜ x 8½. 0-486-63433-7

History of Math

THE WORKS OF ARCHIMEDES, Archimedes (T. L. Heath, ed.). Topics include the famous problems of the ratio of the areas of a cylinder and an inscribed sphere; the measurement of a circle; the properties of conoids, spheroids, and spirals; and the quadrature of the parabola. Informative introduction. clxxxvi+326pp. 5⅜ x 8½. 0-486-42084-1

A SHORT ACCOUNT OF THE HISTORY OF MATHEMATICS, W. W. Rouse Ball. One of clearest, most authoritative surveys from the Egyptians and Phoenicians through 19th-century figures such as Grassman, Galois, Riemann. Fourth edition. 522pp. 5⅜ x 8½. 0-486-20630-0

THE HISTORY OF THE CALCULUS AND ITS CONCEPTUAL DEVELOPMENT, Carl B. Boyer. Origins in antiquity, medieval contributions, work of Newton, Leibniz, rigorous formulation. Treatment is verbal. 346pp. 5⅜ x 8½. 0-486-60509-4

THE HISTORICAL ROOTS OF ELEMENTARY MATHEMATICS, Lucas N. H. Bunt, Phillip S. Jones, and Jack D. Bedient. Fundamental underpinnings of modern arithmetic, algebra, geometry and number systems derived from ancient civilizations. 320pp. 5⅜ x 8½. 0-486-25563-8

A HISTORY OF MATHEMATICAL NOTATIONS, Florian Cajori. This classic study notes the first appearance of a mathematical symbol and its origin, the competition it encountered, its spread among writers in different countries, its rise to popularity, its eventual decline or ultimate survival. Original 1929 two-volume edition presented here in one volume. xxviii+820pp. 5⅜ x 8½. 0-486-67766-4

GAMES, GODS & GAMBLING: A HISTORY OF PROBABILITY AND STATISTICAL IDEAS, F. N. David. Episodes from the lives of Galileo, Fermat, Pascal, and others illustrate this fascinating account of the roots of mathematics. Features thought-provoking references to classics, archaeology, biography, poetry. 1962 edition. 304pp. 5⅜ x 8½. (Available in U.S. only.) 0-486-40023-9

OF MEN AND NUMBERS: THE STORY OF THE GREAT MATHEMATICIANS, Jane Muir. Fascinating accounts of the lives and accomplishments of history's greatest mathematical minds–Pythagoras, Descartes, Euler, Pascal, Cantor, many more. Anecdotal, illuminating. 30 diagrams. Bibliography. 256pp. 5⅜ x 8½. 0-486-28973-7

HISTORY OF MATHEMATICS, David E. Smith. Nontechnical survey from ancient Greece and Orient to late 19th century; evolution of arithmetic, geometry, trigonometry, calculating devices, algebra, the calculus. 362 illustrations. 1,355pp. 5⅜ x 8½. Two-vol. set. Vol. I: 0-486-20429-4 Vol. II: 0-486-20430-8

A CONCISE HISTORY OF MATHEMATICS, Dirk J. Struik. The best brief history of mathematics. Stresses origins and covers every major figure from ancient Near East to 19th century. 41 illustrations. 195pp. 5⅜ x 8½. 0-486-60255-9

Physics

OPTICAL RESONANCE AND TWO-LEVEL ATOMS, L. Allen and J. H. Eberly. Clear, comprehensive introduction to basic principles behind all quantum optical resonance phenomena. 53 illustrations. Preface. Index. 256pp. 5⅜ x 8½. 0-486-65533-4

QUANTUM THEORY, David Bohm. This advanced undergraduate-level text presents the quantum theory in terms of qualitative and imaginative concepts, followed by specific applications worked out in mathematical detail. Preface. Index. 655pp. 5⅜ x 8½. 0-486-65969-0

ATOMIC PHYSICS (8th EDITION), Max Born. Nobel laureate's lucid treatment of kinetic theory of gases, elementary particles, nuclear atom, wave-corpuscles, atomic structure and spectral lines, much more. Over 40 appendices, bibliography. 495pp. 5⅜ x 8½. 0-486-65984-4

A SOPHISTICATE'S PRIMER OF RELATIVITY, P. W. Bridgman. Geared toward readers already acquainted with special relativity, this book transcends the view of theory as a working tool to answer natural questions: What is a frame of reference? What is a "law of nature"? What is the role of the "observer"? Extensive treatment, written in terms accessible to those without a scientific background. 1983 ed. xlviii+172pp. 5⅜ x 8½. 0-486-42549-5

AN INTRODUCTION TO HAMILTONIAN OPTICS, H. A. Buchdahl. Detailed account of the Hamiltonian treatment of aberration theory in geometrical optics. Many classes of optical systems defined in terms of the symmetries they possess. Problems with detailed solutions. 1970 edition. xv + 360pp. 5⅜ x 8½. 0-486-67597-1

PRIMER OF QUANTUM MECHANICS, Marvin Chester. Introductory text examines the classical quantum bead on a track: its state and representations; operator eigenvalues; harmonic oscillator and bound bead in a symmetric force field; and bead in a spherical shell. Other topics include spin, matrices, and the structure of quantum mechanics; the simplest atom; indistinguishable particles; and stationary-state perturbation theory. 1992 ed. xiv+314pp. 6⅛ x 9¼. 0-486-42878-8

LECTURES ON QUANTUM MECHANICS, Paul A. M. Dirac. Four concise, brilliant lectures on mathematical methods in quantum mechanics from Nobel Prize-winning quantum pioneer build on idea of visualizing quantum theory through the use of classical mechanics. 96pp. 5⅜ x 8½. 0-486-41713-1

THIRTY YEARS THAT SHOOK PHYSICS: THE STORY OF QUANTUM THEORY, George Gamow. Lucid, accessible introduction to influential theory of energy and matter. Careful explanations of Dirac's anti-particles, Bohr's model of the atom, much more. 12 plates. Numerous drawings. 240pp. 5⅜ x 8½. 0-486-24895-X

ELECTRONIC STRUCTURE AND THE PROPERTIES OF SOLIDS: THE PHYSICS OF THE CHEMICAL BOND, Walter A. Harrison. Innovative text offers basic understanding of the electronic structure of covalent and ionic solids, simple metals, transition metals and their compounds. Problems. 1980 edition. 582pp. 6⅛ x 9¼. 0-486-66021-4

HYDRODYNAMIC AND HYDROMAGNETIC STABILITY, S. Chandrasekhar. Lucid examination of the Rayleigh-Benard problem; clear coverage of the theory of instabilities causing convection. 704pp. 5⅜ x 8¼. 0-486-64071-X

INVESTIGATIONS ON THE THEORY OF THE BROWNIAN MOVEMENT, Albert Einstein. Five papers (1905–8) investigating dynamics of Brownian motion and evolving elementary theory. Notes by R. Fürth. 122pp. 5⅜ x 8½. 0-486-60304-0

THE PHYSICS OF WAVES, William C. Elmore and Mark A. Heald. Unique overview of classical wave theory. Acoustics, optics, electromagnetic radiation, more. Ideal as classroom text or for self-study. Problems. 477pp. 5⅜ x 8½. 0-486-64926-1

GRAVITY, George Gamow. Distinguished physicist and teacher takes reader-friendly look at three scientists whose work unlocked many of the mysteries behind the laws of physics: Galileo, Newton, and Einstein. Most of the book focuses on Newton's ideas, with a concluding chapter on post-Einsteinian speculations concerning the relationship between gravity and other physical phenomena. 160pp. 5⅜ x 8½. 0-486-42563-0

PHYSICAL PRINCIPLES OF THE QUANTUM THEORY, Werner Heisenberg. Nobel Laureate discusses quantum theory, uncertainty, wave mechanics, work of Dirac, Schroedinger, Compton, Wilson, Einstein, etc. 184pp. 5⅜ x 8½. 0-486-60113-7

ATOMIC SPECTRA AND ATOMIC STRUCTURE, Gerhard Herzberg. One of best introductions; especially for specialist in other fields. Treatment is physical rather than mathematical. 80 illustrations. 257pp. 5⅜ x 8½. 0-486-60115-3

AN INTRODUCTION TO STATISTICAL THERMODYNAMICS, Terrell L. Hill. Excellent basic text offers wide-ranging coverage of quantum statistical mechanics, systems of interacting molecules, quantum statistics, more. 523pp. 5⅜ x 8½. 0-486-65242-4

THEORETICAL PHYSICS, Georg Joos, with Ira M. Freeman. Classic overview covers essential math, mechanics, electromagnetic theory, thermodynamics, quantum mechanics, nuclear physics, other topics. First paperback edition. xxiii + 885pp. 5⅜ x 8½. 0-486-65227-0

PROBLEMS AND SOLUTIONS IN QUANTUM CHEMISTRY AND PHYSICS, Charles S. Johnson, Jr. and Lee G. Pedersen. Unusually varied problems, detailed solutions in coverage of quantum mechanics, wave mechanics, angular momentum, molecular spectroscopy, more. 280 problems plus 139 supplementary exercises. 430pp. 6½ x 9¼. 0-486-65236-X

THEORETICAL SOLID STATE PHYSICS, Vol. 1: Perfect Lattices in Equilibrium; Vol. II: Non-Equilibrium and Disorder, William Jones and Norman H. March. Monumental reference work covers fundamental theory of equilibrium properties of perfect crystalline solids, non-equilibrium properties, defects and disordered systems. Appendices. Problems. Preface. Diagrams. Index. Bibliography. Total of 1,301pp. 5⅜ x 8½. Two volumes. Vol. I: 0-486-65015-4 Vol. II: 0-486-65016-2

WHAT IS RELATIVITY? L. D. Landau and G. B. Rumer. Written by a Nobel Prize physicist and his distinguished colleague, this compelling book explains the special theory of relativity to readers with no scientific background, using such familiar objects as trains, rulers, and clocks. 1960 ed. vi+72pp. 5⅜ x 8½. 0-486-42806-0